# THE
# SCIENCE
# OF
# ASTRONOMY

# THE SCIENCE OF ASTRONOMY

**HARPER & ROW, PUBLISHERS**
New York   Evanston   San Francisco   London

Sponsoring Editor: John A. Woods
Project Editor: Duncan R. Hazard
Designer: T. R. Funderburk
Production Supervisor: Stefania J. Taflinska

THE SCIENCE OF ASTRONOMY

Library of Congress Cataloging in Publication Data

Main entry under title:

The Science of astronomy.

    1.    Astronomy.
QB43.2.S37       520        73-10684
ISBN 0-06-041446-4

# CONTENTS

# PREFACE

*The Science of Astronomy* is designed for use in the one-semester introductory course as taught in colleges around the country. The book is almost purely descriptive in nature, requiring little or no mathematical background. It presents the concepts of astronomy in a straightforward, logical manner with many line drawings and photographs.

The text begins by giving the student a background in the concepts necessary to understand the subject—the visible sky and light. It then examines the solar system with chapters on the moon, planets, asteroids, the sun, and finally, the stars and galaxies, covering their nature and cosmology in detail.

In an easy-to-read fashion, *The Science of Astronomy* is designed to give the nonscience major a feel for the subject. It is not the aim of this book to go into detail in any particular area of astronomy. When the student finishes the book, he should have an intuitive feel for the subject and be able to understand the importance of scientific probing of the solar system and the universe.

This book is the joint effort of several people. It was initiated by Professor Harry Crull of The State University of New York at Albany. Several instructors have commented on it, and Dr. William Kaufmann, Director of the Griffith Observatory, was actively involved in the final writing of the manuscript.

# THE
# SCIENCE
# OF
# ASTRONOMY

# CHAPTER 1
# THE LURE OF ASTRONOMY

Astronomy has the dual distinction of being the oldest science as well as one of the most fascinating fields of modern research. In the mid-twentieth century many scientists, both young and old, turned their attention to a variety of astronomical and astrophysical problems. As a result of their efforts, we now seem to be on the brink of a number of fundamental changes in our thinking about the universe. Astronomy has challenged man's ingenuity from earliest times. It has become even more challenging now, at the beginning of the age of space.

## 1.1 THE NEED FOR BASIC RESEARCH

The demands on our attention made by our modern and complex society exceed in volume and intensity anything that our parents or grandparents could possibly have imagined. From all sides and through all media our senses are bombarded with assertions of the overwhelming importance and the absolute priority of each facet of our civilization. This is partly a result of the inherent complexity of modern life, and partly a result of the expansion of facilities for rapid communication of ideas. Within little more than a generation, we have moved from an era in which a trip around the world in 80 days was almost unthinkable to one in which astronauts circle the globe in little more than 80 minutes. A century ago a trip across the continental United States constituted the adventure of a lifetime. Today we can photograph continents from space (Figure 1.1). Along with our geographic horizons, our mental horizons and the problems competing for our attention have expanded. Twentieth-century man has seen vastly more technical progress than all the previous generations of mankind. Everyday life at the end of the eighteenth century resembled that of the Roman empire more than that of the 1970s. Knowledge has expanded exponentially, and technology has followed in its footsteps. However, our awareness of unsolved problems has increased even more rapidly than our solution of previously recognized problems.

Each increase in our knowledge, each fact discovered, each problem solved has brought with it a host of questions. In a way we are almost like the characters in *Through the Looking-Glass*, who must run as rapidly as possible to stay where they are. New decisions are forced upon us every day. At the same time, we must prepare ourselves to answer questions that may confront us in the future.

In the exploration of space, as in the equally challenging exploration of our own planetary environment, this preparation takes the form of basic research. Knowledge of the universe is an essential prerequisite to intelligent and effective decision making in a technologically oriented society that contemplates further space exploration in the years to come.

**Figure 1.1 The Earth from Space**
This photograph of the earth was taken during the Apollo 10 mission by Astronauts Stafford, Young, and Cernan. The west coast of North America, especially Baha, California, is clearly visible. Clouds obscure most of the rest of the landmass. NASA photograph.

## 1.2 MAN, THE CURIOUS CREATURE

Throughout history and in the legends of prehistory, man was always a curious creature, investigating and trying to explain his surroundings. Yet, despite this fact, mankind remained in ignorance of the basic nature of the physical universe for the greater part of its existence. Primitive man explained events in the world by populating the earth and sky with capricious spirits, some evil and some benign. Even the seemingly fixed stars became the outline of mythological beasts and folk heroes.

3

Nevertheless, it was early in the development of civilization that men drew order from the apparent chaos of the sky. Observation of the phases of the moon gave rise to surprisingly accurate calendars, primitive navigation was born, and observatories of astonishing sophistication appeared. In Great Britain, the megaliths of Stonehenge were constructed, perhaps, to facilitate astronomical observations.

Astronomy has had an impact on many facets of life for centuries. Authors, poets, artists, sculptors, and musicians have drawn inspiration from the heavens. Cosmology has been a vibrant field of study from the time of Genesis. Discovery and exploration on earth and in space have been guided by the stars from earliest times. Physics and chemistry have benefited from a study of the behavior and properties of matter under conditions impossible to duplicate on earth.

Some of these developments have been slow, others quick. In the early years of the last century an American president, Thomas Jefferson, said that he would rather believe that a group of Yankee professors were liars than to believe that stones fell from the sky. Today, less than two centuries later, museums are full of meteorites that have fallen from the sky. Today there are still many who use the position of the crescent moon in the western sky to predict (or pretend to predict) the weather, not realizing that this position depends entirely on the season of the year. At the corresponding season, year after year, century after century, the crescent moon has been in exactly the same position. Its relationship to the weather is not one of causability; on the contrary, both the rainy season and the position of the moon are the simultaneous result of a single astronomical phenomenon. Countless examples of such misinterpretations of the very simplest of physical phenomena could be given.

Yet however falteringly man has moved forward, however frequently he has erred, however many blunders he has made, his curiosity about why the world is the way it is has led him along intellectual paths untrodden by the other species that populate the earth. No ape, as far as we know, has ever wondered why grass is green, why the sky is blue, why the sun rises and sets, why the tides rise and fall. Only man has lifted his eyes above the surface of the earth to the beauty of the nighttime sky and wondered who or what lit the candles of the stars.

## 1.3 THE FUTURE OF ASTRONOMY

The history of astronomy is both long and fascinating, yet the future promises to be even more exciting. Now, as the twentieth century draws toward its close, we seem to be on the threshold of answers to a host of questions regarding the nature of the universe. In the chapters that follow, we shall explore some of these questions.

Probably the most exciting area of contemporary astronomy is that of space exploration and travel. Ever since the successful voyage of Sputnik I, our eyes have turned with new enthusiasm to the heavens. Each small step forward, from manned orbital flights to space walks, docking, the performance in space of what would be simple tasks on the earth's surface, prolonged exposure to weightlessness, unmanned lunar probes and orbiters, Martian probes, manned lunar orbiters, and finally man's first step on the moon, reads like a lexicon of the science fiction of two decades ago.

A dream of centuries was realized when the first man stepped on the surface of the moon (Figure 1.2). The most costly pieces of material in all the history of mankind were the samples of lunar rocks returned by the astronauts. Overshadowed by the spectacular success of Apollo 11 in 1969 (Figure 1.3), but important nonetheless, were the brilliant exploratory vehicles that visited the vicinity of the planet Mars and sent back magnificent photographs of its surface. Plans for the future of space exploration are breathtaking, and the knowledge gained from their realization may exceed all expectations. Our position is like that of the people who stood on a dock one summer day in 1492 and watched three little ships sail westward across an unknown ocean. Those who today decry the expense of the lunar

Figure 1.2 **The First Footprint on the Moon**
On July 20, 1969, Astronaut Neil Armstrong stepped on the surface of the moon. He was the first man to tread a celestial body other than the earth. NASA photograph.

voyages might do well to remember the words of Senator Daniel Webster, who, in 1840, informed the Congress of the United States that he would not vote one cent for the development of the West, which would forever be a "howling wilderness."

Another, less publicized, area of great activity and interest in modern astronomy is interstellar chemistry. Until quite recently, little was known or even suspected about the combination of atoms into molecules in interstellar space. The atmospheres of the cooler stars have long been known to contain very simple molecules. Recently, intensive study has resulted in the discovery of compounds and even chemical radicals in space. Interestingly enough, the elements frequently involved have been those which, in part at least,

Figure 1.3 **The Apollo 11 Lunar Module**
The module is ascending to dock with the command module. The earth is rising in the background. NASA photograph.

constitute the foundations of life—hydrogen, oxygen, carbon, and nitrogen.

Rapid advances are also being made in radio astronomy. Originally, the investigation of radio waves coming to us from space was hampered by the astronomer's inability to resolve and identify the optical equivalent of radio sources because of primitive equipment

7

**Figure 1.4 The Arecibo Radio Telescope**
This is the largest radio telescope in the world. The dish is 1000 feet in diameter
and is installed in a bowl-shaped valley in Puerto Rico. The telescope itself is not
steerable, but moving the receptor suspended 435 feet above the reflector 20° in
any direction from the zenith permits astronomers to survey a belt around the sky
40° wide. The Arecibo National Astronomy and Ionosphere Center is operated by
Cornell University under contract to the National Science Foundation.

and the long wavelength of the radio radiation. Recent and impend-
ing improvements in instrumentation promise to remove many, if
not all, of these disabilities. Indeed, a considerable portion of our
present knowledge of the interstellar chemistry discussed in the preceding paragraph has come from the work of the radio astronomer.
Long-baseline interferometry, which uses several radio telescopes at
the same time, provides us with amazingly sharp pictures of the
radio sky, far surpassing those produced by optical instruments, and
opens entire new vistas in this most exciting area. Figure 1.4 shows
the Arecibo radio telescope in Puerto Rico, the largest radio telescope in the world.

Closely related to (and first detected by) radio techniques are

Figure 1.5 **The Crab Nebula**
This nebula is all that is left of a supernova observed in 1054. The central body is a pulsar. Photograph from the Hale Observatories.

the quasars. In the early 1960s, astronomers optically identified the first of what are now known as *quasars*—strong radio sources that appeared optically as faint stars which seemed to be traveling away from us at speeds close to the speed of light. The most distant as well as the intrinsically most luminous class of objects known, they remain today among the most mysterious objects on the frontiers of astronomy.

Still another object, even newer to astronomers, is the *pulsar* (Figure 1.5), of which some 60 are recognized. The energy output of these stars varies rapidly and with incredible precision. Today, pulsars are thought to be very dense stars resulting from a stellar explosion known as supernova. Their variation is probably caused by their rapid rotation after that catastrophic event.

More speculative even than pulsars are the recently postulated *collapsars*, or *black holes*. These are hypothetical stars of such enormously high density (and hence with such huge gravitational fields) that even light cannot attain the velocity needed to escape them. These objects cannot be seen or detected in any way, except perhaps by their gravitational effect on nearby objects.

Another area that we shall consider in this book is *cosmology*, the study of the nature, evolution, governing laws, and probable future of the universe. It is certainly the most inclusive and widespread investigation in which man can participate.

## 1.4 THE PERILS OF PREDICTION

Although men have made giant steps forward, standing each time on the shoulders of those who preceded them, those who have sought, as we are doing here, to point out the future have all too often made mistakes that have become classics in the history of science.

In 1932, an eminent American astronomer said, "There is no hope for the fanciful idea of reaching the moon." That same year, another prominent scientist said, "Anyone who looks for a source of power in the transformation of atoms is talking moonshine." Some years later, but before Alamagordo, an admiral in the U.S. Navy said, "The A bomb is the biggest fool thing we have ever done. [It] will never go off, and I speak as an expert on explosives." During the 1930s, the Astronomer Royal of the British Empire described the idea of space travel as "pure bilge."

Perhaps, in the not-too-distant future, our own ideas may be regarded as little more than stumbles in the dark. It is only with this chastening thought in mind that we can present the conclusions man has thus far drawn from what is truly the most all-encompassing intellectual activity of the human race, the contemplation of the universe.

# CHAPTER 2
# THE VISIBLE SKY

Celestial observation can take many forms, ranging from casual glances at the nighttime sky to sophisticated studies of far galaxies with the finest modern equipment. In order to examine any faint object with his telescope, an astronomer obviously must have some way of knowing exactly where to point his instrument. Star catalogs and systems of coordinates for locating celestial objects accurately are therefore essential and will be our first subject of discussion.

   In addition to star catalogs and coordinate systems, we must discuss the subject of time. At any one instant, as we stand at different locations on the earth, there are different stars in the sky. An understanding of time and timekeeping is therefore relevant to the problem of knowing where to point the astronomer's telescope.

## 2.1 THE ANCIENT CONSTELLATIONS

The word *constellation,* in the ancient sense, meant "a conventional or historically accepted group of stars in the sky." Ancient man saw in these groups of stars the outlines of beasts and men, much as we today sometimes see "faces" in the clouds. The poet Aratus, in 270 B.C., described the mythical creatures in the heavens and told stories about them, some of which have come down to the present day. However, the oldest mentions of the star figures are from millennia before the time of Christ, and their origins are lost in antiquity. Probably, they originated somewhere in the Tigris-Euphrates valley and spread with civilization. The oldest star catalog we possess was compiled about A.D. 150 by the astronomer Ptolemy, who listed the names of bright stars and described their location.

The ancients identified 48 constellations, or groups of stars. However, this still left many unnamed areas in the sky where bright stars or stars of medium brilliance did not appear. Nor did the old constellations cover the southern sky, since European civilization developed in the Northern Hemisphere.

## 2.2 THE MODERN CONSTELLATIONS

As time went on, the concept of constellations gradually changed, until today we think of them not as groups of stars, but as areas of the sky. There are 88 modern constellations, and they cover the sky completely. Originally, the boundaries of even the modern constellations wandered rather aimlessly among the stars, enclosing the star groups named by the ancients. Indeed, star maps published in the early part of the twentieth century show boundaries meandering about to enclose the conventional figures. It was not until 1928 that the International Astronomical Union established the boundaries that are accepted today. Modern sky maps generally do not show the old figures of beasts and men that the ancients associated with the constellations, but they do retain the old names. We also retain the proper names of the bright stars that have come down to us

**Figure 2.1  Orion in Bayer's Atlas**
The constellation of Orion as depicted in Johann Bayer's *Uranometria*, Ulmae
1661. Dudley Observatory photograph.

from antiquity. Even in highly sophisticated scientific journals, these
stars are referred to by their ancient names.

## 2.3 DESIGNATIONS OF STARS
##    WITHIN CONSTELLATIONS

With the invention of the telescope in the seventeenth century, it
became obvious that there were many millions of stars in the sky.
Even before Galileo, however, it was apparent that not every star
could receive a proper name. Nonetheless, the derivation of the an-

13

cient proper names is interesting. The star Betelgeuse, for example, in the upper left corner of the constellation of Orion (Figure 2.1), is in the shoulder of the conventional figure. The meaning of the word "Betelgeuse" is uncertain, but most authorities agree that it translates best from Arabic as "the armpit of the central one," although we might call it the shoulder of the giant. The derivation of other names, such as Polaris (for the north pole star), is more obvious.

As the study of astronomy progressed, astronomers needed a simpler and more universal method of designating stars. In 1603, Johann Bayer suggested in his atlas a method of designating the brighter stars that is still in use. He gave Greek letters, in alphabetical order, to the stars of each constellation. Following the Greek letter he used the Latin possessive form of the name of the constellation. In general, he started with the brightest star, which he called alpha; the second brightest star was called beta, and so on. The system was not absolutely inviolable, however. Sometimes a group of stars of approximately the same brightness were lettered according to their position, moving across the constellation. For example, the star Betelgeuse in the constellation of Orion (Figures 2.2 and 2.3) is referred to as Alpha Orionis, whereas Rigel, actually a slightly brighter star in the same constellation (though it may not appear so to the naked eye), is called Beta Orionis.

As investigations in astronomy proceeded, it became obvious that the Greek alphabet was not large enough to provide designations for all the stars. Approximately a century after Bayer, the English astronomer John Flamsteed suggested a more extensive system. After the Greek letters had been exhausted, the fainter stars were numbered, moving across the constellations from east to west. This system included only stars that were visible to the naked eye. Today, large telescopes far beyond the dreams of ancient and even early modern astronomers provide photographs of literally millions of stars, all of which must be designated in some way. Most faint stars today are designated by running numbers in catalogs, which we shall discuss in the next section.

14

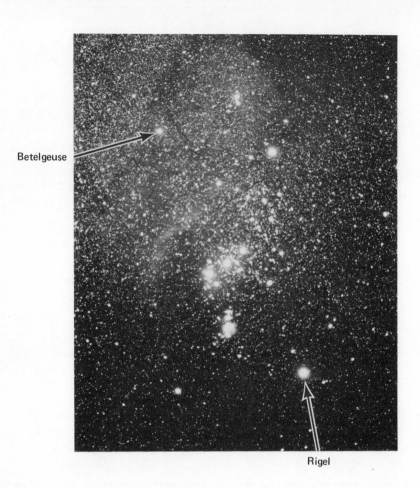

Betelgeuse

Rigel

Figure 2.2 **The Region of Orion**
Note that Rigel ($\beta$ Orionis), a blue star, photographs brighter than Betelgeuse ($\alpha$ Orionis); yet to the eye Betelgeuse appears slightly brighter. (Compare with Figures 2.1 and 2.3.)

Students who wish to familiarize themselves with the constellations and the names of the bright stars should consult the star charts in the appendix. Occasionally, a bright object will be seen in the heavens that does not coincide with anything plotted on the star map. The moon and the five planets easily visible to the unaided eye at various times of the year are not plotted on star maps because they do not occupy permanent places. The word "planet" means "wanderer," and the planets do wander about, or change their position against the fixed pattern of the sky. To identify bright planets,

**15**

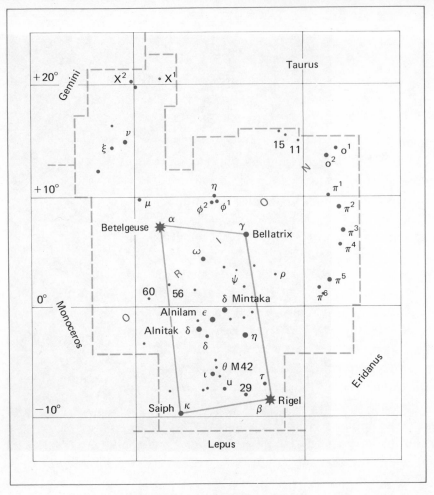

Figure 2.3 **The Constellation of Orion**
This diagram gives the I.A.U. boundary and the Greek letter names of the brighter stars. (Compare with Figure 2.2.)

one must consult publications giving their positions during the year, such as *Sky and Telescope*, the *Griffith Observer*, or the *Observers' Handbook* (published by the Royal Astronomical Society of Canada). Venus and Jupiter are particularly conspicuous at times. Saturn and Mars are less so, and Mercury is only occasionally noticeable. Not to be overlooked as an aid to the study of constellations are the excellent planetarium lectures given at various institutions throughout the country.

**16**

## 2.4 STAR CATALOGS

Although studying the constellations and learning the names of bright stars so as to be able to identify them on a dark, clear night is interesting for the amateur astronomer, it really has no place in research astronomy. There are about 100 billion stars in our galaxy, many of which cannot be seen even with the best telescopes. However, the number of stars that can be observed and studied is certainly prodigious. Obviously, the oldest catalogs list relatively few stars. Ptolemy's *Almagest* gives the position of 1025 stars. Catalogs were also compiled by Ulug Beg in 1450 and Tycho Brahe in 1580. With the invention and continued refinement of the telescope, the number of observable stars grew by leaps and bounds.

At present there are various methods of designating stars and locating them for restudy at a later date. The first type of star catalog is called a *Durchmusterung* (literally, a listing, checking, or "muster" of parts). In the construction of such a catalog, a *limiting brightness* is chosen. The fainter the limit, the more stars there will be in the catalog and the larger it will be. If the catalog is made in the Northern Hemisphere, the southern circumpolar regions will be omitted. After the limiting brightness has been chosen, each star is measured and, usually, a chart is made. The *Bonner Durchmusterung* (made at Bonn, Germany, about 1862), for example, has large charts. A typical star designation in this catalog might be "BD +5° 2173." This simply means that the star is the 2173rd star, counting eastward from a point in the sky called the vernal equinox in a band 1° wide (from 5° to 6°) in the northern half of the sky. Many such star catalogs have been made, and the degree of precision of the coordinates is usually high.

Another type of catalog is the *precision catalog*, which may be fundamental or differential. The *fundamental precision catalog* includes the position of a certain number of fundamental stars (stars from which the positions of other objects can be measured) on a given date. These catalogs may contain only a few thousand stars, but they are analogous to the benchmarks of a surveyor and provide reference points from which other catalogs can be constructed. A *differential precision catalog* uses the star positions from the fundamental catalog as standards in terms of which the positions of other stars can be stated.

There are also special catalogs such as those listing double stars, variable stars, or nebulae. For example, the *Yale Catalog* of bright stars contains approximately 10,000 stars and includes a wide range of information on each star in addition to its position. Since its publication in 1930, this catalog has been revised several times to include more up-to-date data.

Of particular interest is a catalog of 103 *nebulas,* or cloudy objects, compiled in 1784 by Messier, a French comet-seeker. The numbers in this catalog, preceded by the letter **M**, are used to designate bright "nebulas." For example, the globular cluster of stars in Hercules is M13. (The word "nebula" is used loosely here; a more rigorous definition will be given later.) A more exhaustive catalog of these hazy, nonstellar objects is Dreyer's *New General Catalogue* (NGC) of 1887 and the later *Index Catalogue* (IC). Together these list more than 13,000 objects. The Hercules cluster, M13, is also known as NGC 6205. These designations will be used throughout the text.

In addition to the catalogs we have mentioned, there are annual publications such as the *American Ephemeris and Nautical Almanac* and the *Apparent Places of the Fundamental Stars*, which also provide lists of the positions of the stars and other objects in the sky. Strictly speaking, however, these publications are not catalogs in the true sense.

Accompanying many of the catalogs are *stellar atlases*, which are simply maps of the sky. The early ones were made by plotting dots for the stars and were constructed from telescopic observations. The size of the dot indicated the brilliance, or magnitude, of the stars. More recent maps are made from photographs. The most elaborate and complete of all the stellar maps is the *National Geographic Palomar Sky Survey*, which covers virtually all the sky that

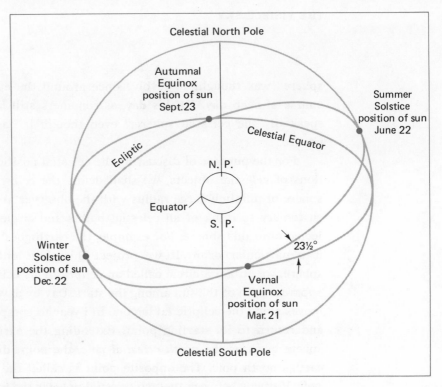

**Figure 2.4  The Celestial Sphere**
All the astronomical objects we see in the sky may be thought of as located on an extremely large sphere, with the earth at its center. Projecting the earth's poles and equator outward gives us their celestial counterparts. The apparent path of the sun among the stars is called the ecliptic. The ecliptic crosses the celestial equator at two points, called the vernal equinox and the autumnal equinox.

can be seen from the Mt. Palomar Observatory in California. It includes stars one-millionth as bright as those visible to the unaided eye and sells for about $2000.

## 2.5 THE CELESTIAL SPHERE

As ancient man looked up into the heavens and sought order in the Universe, he noticed that the "fixed stars" rose and set like the sun, moon, and planets. Unlike these bodies, however, the stars did not change their positions relative to one another. Ancient man therefore assumed that the stars were attached to the inside of a huge sphere with the earth at its center (Figure 2.4). This "celestial sphere" was thought to revolve once around the earth during the course of each day. To this day, astronomers still find it useful to speak of "the celestial sphere," even though it has no actual existence.

For the purpose of discussing the celestial positions and the motions of celestial objects, we shall define the *celestial sphere* as a sphere of infinitely large radius with the observer at its center. It is customary to think of all celestial points and circles as being projected onto this sphere. For example, the earth goes around the sun in a particular orbit. If we project this orbit onto the celestial sphere, we obtain a circle called an *ecliptic*. This ecliptic is also the apparent path of the sun among the stars. Day by day, the sun slowly moves along the ecliptic taking one full year to complete its journey and return to its starting point. Extending the earth's axis to the sphere gives us the *north celestial pole*, the point directly over the earth's north pole. The opposite point is called the *south celestial pole*. Halfway between the two celestial poles is the projection of the earth's equator, called the *celestial equator.*

A *great circle* is the largest circle that can be drawn on a sphere. Strictly speaking, a great circle is a circle whose plane passes through the center of the sphere. All other possible circles are called *small circles*. The ecliptic and the celestial equator are great circles on the celestial sphere.

Two great circles upon the surface of a sphere intersect at two opposite points. In the case of the celestial equator and the ecliptic, these two points are the *vernal equinox* and the *autumnal equinox*.

With this material as background, we may now proceed to devise *coordinate systems* for locating objects on the celestial sphere.

## 2.6 COORDINATE SYSTEMS

We are all familiar with the process of finding a house in a strange neighborhood by using its street address. Every street-numbering

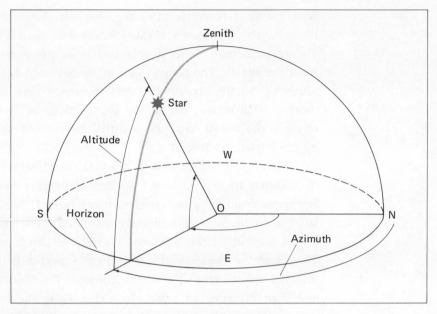

**Figure 2.5  The Horizon System**
In this system the observer is located at point O, and the visible half of the celestial sphere appears as a hemisphere above him. The points of the compass (N, E, S, W) and the zenith are as indicated. The azimuth and altitude of an object on the celestial sphere are measured as shown in the diagram.

system is basically a coordinate system of some sort. The most precise coordinate system for the surface of the earth is that of *latitude* and *longitude*.

The *latitude* of any point on the earth is its angular distance, in degrees, minutes, and seconds, north or south of the earth's equator. Its *longitude* is its distance, in the same units, east or west of the Greenwich meridian. If the two coordinates are known, the location of any feature of the earth's surface can be given. Similar systems have been developed for the sky, using points and lines on the celestial sphere. We shall consider four of these systems.

The first is the *horizon system* (Figure 2.5). It is the simplest system, but has notable disadvantages. The basic point in the horizon system is the zenith. The *zenith* is that point in the sky which is directly overhead of the observer. As the earth turns upon its axis, the observer's zenith moves eastward among the stars. The fundamental circle of this system is the *horizon*, a circle 90° from the zenith. The term "horizon" is not used here in the popular sense, as the line, frequently very uneven, where the sky seems to meet the landscape, but in the technical sense, meaning the great circle in the sky 90° from the zenith (and most closely approximated by the horizon visible at sea).

The *altitude* of an object, the first coordinate in this system, is its distance in degrees and fractions of degrees above the horizon. It ranges from zero (when the object is just rising or setting) to 90° (when the object is directly overhead). The second coordinate of the horizon is the *azimuth*. This is the direction of the object. To determine the azimuth, we draw a perpendicular line on the celestial sphere from the zenith through the object to the horizon and then measure an angle clockwise along the horizon from the north point to the foot of the perpendicular. The zero point used by astronomers is the north point. At any given instant, a particular object has a unique altitude and azimuth for a particular observer. Different observers having different zeniths and horizons would, at a particular instant, give corresponding different altitudes and azimuths for the same object. Even for the same observer, the altitude and azimuth of an object changes with time. Consequently, the horizon system is strictly a local system.

Another system, based on the rotation of the earth, is the *equatorial system* (Figure 2.6). If we think of the earth as a small point at the center of the celestial sphere and expand the equatorial latitude and longitude coordinate system until it reaches this sphere, we have what is called the *celestial equatorial system* of coordinates. The fundamental points are the north and south celestial poles, directly above the north and south poles of the earth. The fundamental circle is the celestial equator, halfway between the two poles and directly above the earth's equator. The first coordinate, corresponding to terrestrial latitude, is the coordinate of *declination*,

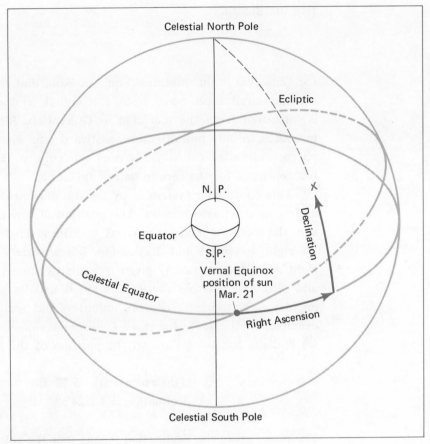

**Figure 2.6  The Equatorial System**
The earth is at the center of the celestial sphere. The right ascension of an object on the sphere is measured along the celestial equator eastward from the vernal equinox. The declination is measured north or south of the celestial equator, as shown in the diagram. This system is widely used among astronomers. Right ascension and declination on the celestial sphere are analogous to longitude and latitude on the earth.

the angular distance measured north or south of the celestial equator along a great circle passing through the celestial poles. Just as there are many places on the surface of the earth with the same latitude, so there may be many stars with the same declination. The second coordinate, corresponding to longitude on the surface of the earth, is called *right ascension*. Right ascension is the angular distance from the vernal equinox, measured eastward through either 360° or 24 hours. Opposite the vernal equinox, in right ascension 180°

(or 12 hours) is the autumnal equinox. Note that right ascension is *not* measured both ways from the vernal equinox as longitude is measured from the meridian at Greenwich. Right ascension is measured in only one direction, eastward (the direction of the apparent motion of the sun). It does not stop at 180° or 12 hours, but continues for the second half of the circle.

This coordinate system is by far the most widely used system among modern astronomers. The position of any object, be it the sun, the moon, a planet, or a star, can be given by two numbers, it's right ascension and declination. For a variety of reasons, the right ascension is usually given in units of time (hours, minutes, and seconds), whereas the declination is usually given in units of angular measurement (degrees, minutes, and seconds). Thus, for example, in the 1950 *American Ephemeris and Nautical Almanac*, we find the following data on the position of the star Betelgeuse:

$$\text{right ascension} = 5^h52^m29^s$$
$$\text{declination} = +7°24'5''$$

The plus sign in the declination means that the star is north of the celestial equator. A minus sign is used to indicate the declination of objects in the southern half of the sky.

A third system of celestial coordinates, older but less used, is called the *ecliptic system* (Figure 2.7). The ecliptic system is the same as the equatorial system, except for the fundamental points and circle. The fundamental circle in this case is, as one might expect, the ecliptic. The zero point for measuring celestial longitude, which corresponds to right ascension in the equatorial system, is the vernal equinox. The ecliptic system is sometimes used for locating the sun (whose celestial latitude is always zero), the moon, and the planets. Since the moon and most of the planets have orbits inclined at small angles to the plane of the earth's orbit, they are never seen very far from the ecliptic. The apparent paths of all the principal planets (except Pluto), in fact, lie within a belt some 16°

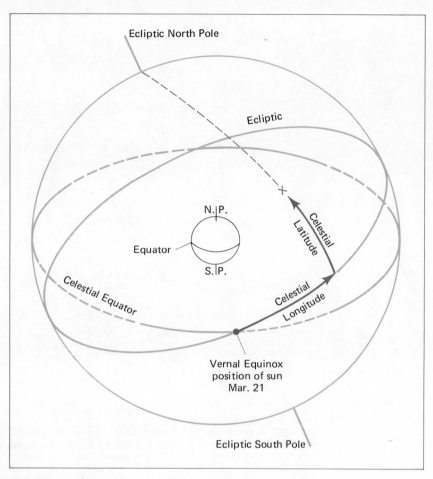

Figure 2.7  **The Ecliptic System**
The earth is at the center of the celestial sphere. Celestial longitude is measured along the ecliptic eastward from the vernal equinox. Celestial latitude is measured north or south of the ecliptic, as shown in the diagram. This system is infrequently used by astronomers.

wide, with the ecliptic as the center line. This region was called the *zodiac* by the ancients, who divided it into 12 constellations.

A fourth system of coordinates, called the *galactic system* (Figure 2.8), is used in studying our galaxy, or the Milky Way. Its fundamental plane is the great circle nearest to the central line of the Milky Way. *Galactic latitude* is the distance in degrees north or south of this particular circle. At present, *galactic longitude* is measured from a point along the galactic equator near the pre-

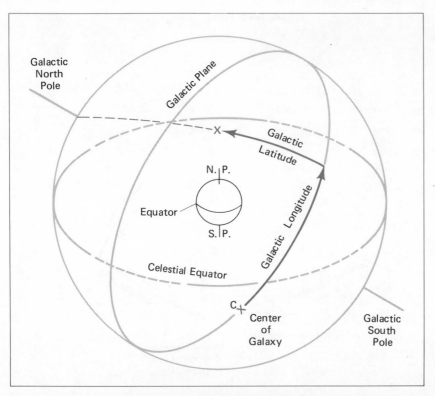

Figure 2.8 **The Galactic System**
The earth is at the center of the celestial sphere. The galactic plane is the great
circle of the sky that most nearly coincides with the central plane of the Milky Way.
This circle is inclined over 60° to the celestial equator. Point C is in the direction
of the center of our galaxy. Its position is given by the coordinates R.A. = 17$^h$42$^m$,
Decl. = −28°55′. Galactic longitude is measured eastward from the galactic center
along the galactic plane. Galactic latitude is measured north and south of the
galactic plane, as shown in the diagram.

sumed direction of the galactic center. This zero point is in the
constellation of Sagittarius. Galactic longitude is measured, in
degrees, in the same direction as right ascension.

It is by means of these systems of coordinates (and methods of
transforming coordinates from one system to another) that an
astronomer is able to locate objects precisely in the sky. The
horizon system is used mostly by surveyors and navigators. The
equatorial system is the system upon which the star catalogs are
based. (The right ascension and declination of many objects for
each hour and each day in the year are given in the *American
Ephemeris and Nautical Almanac*. Transformations between the

horizon system and the equatorial system are easy to make.) The ecliptic system is used in studying the solar system and its members. (Ecliptic coordinates of some members of the solar system are given in the *American Ephemeris*.) Finally, galactic coordinates are especially useful to astronomers studying the structure of the Milky Way. Each system thus serves a purpose, and each is valuable in its own way.

## 2.7 THE SEASONS

Having examined the celestial sphere, we can easily understand why we have seasons here on the earth. As we have mentioned, the ecliptic is the apparent path of the sun against the background of the fixed stars. The celestial equator, on the other hand, is obtained simply by extending the earth's equator out into space. The plane of the ecliptic and the plane of the celestial equator are inclined toward each other by 23½° (Figure 2.9). This is because the earth's axis of rotation is not exactly perpendicular to the plane of the earth's orbit, but is inclined at an angle of 23½°. As a result, the sun spends half the year north of the celestial equator and half the year south of the celestial equator.

As we noted earlier, the ecliptic and the celestial equator intersect at two points, the vernal equinox and the autumnal equinox. At a particular instant each year (usually on March 21), the motion of the earth in its orbit causes the sun to appear to move north across the celestial equator. The point where this occurs is the vernal equinox. Directly opposite the vernal equinox is the autumnal equinox, the point at which the sun moves from the northern half of the sky to the southern half, again crossing the equator. The usual date of the autumnal equinox is September 23. The two points midway between the vernal equinox and the autumnal equinox, at which the distance of the sun from the equator is greatest, are called the summer and winter *solstices*.

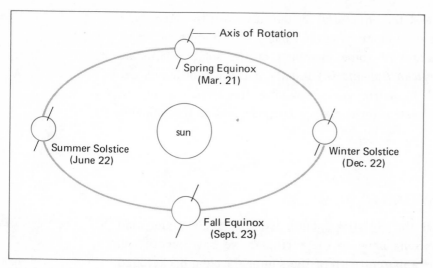

Figure 2.9 **The Tilt of the Earth's Axis**
The earth's axis of rotation points in the same direction in space throughout the year. This direction is tilted 23½° from a line perpendicular to the earth's orbit. As a result, the celestial equator and the ecliptic are inclined toward each other by an angle of 23½°. Thus the sun spends half the year north of the celestial equator and half the year south of the celestial equator.

If there were no inclination between the ecliptic and the celestial equator, the earth would be in a position in relation to the sun equivalent to a perpetual vernal equinox, and we would enjoy an eternal spring. Two effects of the inclination produce the seasons. The first is the length of the day. Because the sun travels a longer path across the sky in June in the Northern Hemisphere, the day is longer than average near the summer solstice (Figure 2.10). For example, at latitude 40°N on June 21, daylight lasts about 15 hours. At the equinox, daylight lasts just over 12 hours, and at the winter solstice in December it lasts only about 9 hours and 20 minutes. Consequently, in the Northern Hemisphere we receive the heat of the sun for a longer period in the months of June and July than we do in December.

Not only is the day longer during the summertime, but the sun is higher in the sky. The noonday altitude of the sun at a point

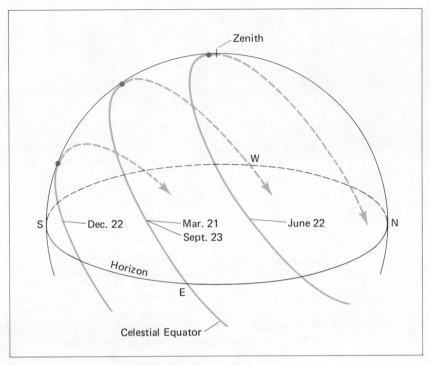

Figure 2.10 **The Length of the Day in the North Temperate Zone**
The hemisphere shown is the half of the celestial sphere above the horizon. The
path of the sun depends on the solar declination. At a latitude of 40°, the
length of the path above the horizon causes daylight to vary from about 15 hours
in June to 9 hours in December.

on earth 40° north of the equator varies from $73\frac{1}{2}°$ at the time of
the summer solstice to $26\frac{1}{2}°$ in the wintertime. Thus a beam of
sunshine 1 mile wide from north to south is greatly spread out in
December and very concentrated in June (Figure 2.11). Indeed,
during June the ray heats a space about 1.04 miles wide from north
to south; in December it must heat a space over $2\frac{1}{4}$ times as great.
Not only is the sun above the horizon nearly 50 percent longer in
the summer, but it is over twice as effective as a source of radiant
energy. This is because the sunshine is spread over less of the earth's
surface.

When the sun is high in the sky in the Northern Hemisphere,
it is low in the heavens in the Southern Hemisphere. Hence the
seasons of the year are reversed in the two hemispheres. Also, in
the Northern Hemisphere we are closer to the sun during the

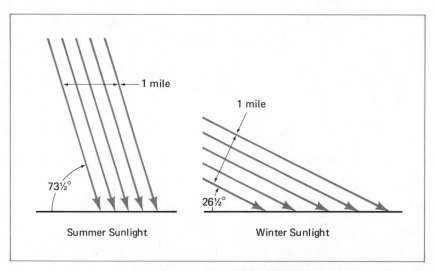

**Figure 2.11  The Spread of Sunlight in Summer and Winter**
At noon on June 21 in latitude 30°N a ray of sunlight has a noonday altitude of
73½° and is spread on the ground by 4 percent to cover 1.04 times its own width.
On December 23 the ray comes from an altitude of 26½° and must heat an area of
ground about 2¼ times as great. Hence the weather is warmer in June than in
December.

winter than we are during the summer. Although the difference is
only about 3 percent, there may be some attenuation of the extremes of the seasons as a result. However, since the Southern
Hemisphere is mostly covered with water, there is, as one would
expect, a tampering effect on the seasonal extremes there. There is no
appreciable systematic difference in the seasons in the two hemispheres.

Another interesting phenomenon associated with the inclination
of the axis of the earth is the midnight sun. If a man is within the
Arctic Circle (that is, at a latitude 66½° or more north of the
equator) on June 21, he will be far enough north so that he can see
past the North Pole to the sun, even at midnight; in other words,
from his point of view, the sun will circle the heavens without
setting. The midnight sun is seen in the Southern Hemisphere on
December 23.

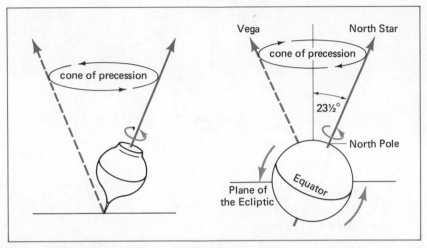

Figure 2.12 **Precession**
The pull of gravity *downward* causes the spinning top to precess (i.e., to change
the direction of its axis with a slow conical motion). The pull of the sun and
moon (and planets) on the earth's equatorial bulge attempts to pull the axis
*upward* to form an angle of 90° with the ecliptic. Note that the resultant precessional
motion is reversed.

## 2.8 PRECESSION

It is often thought that the direction of the earth's axis is absolutely constant in space, unchanging as the earth moves around
its orbit. This is not precisely true. The earth's axis has a very
slow conical motion, much like the axis of a spinning top when
the top is tilted over (Figure 2.12). This motion is called *precession*.

It is well known that the earth is not perfectly spherical.
Rather, it bulges slightly around the equator. This "equatorial
bulge" is produced by the rotation of the earth and is gravitationally
attracted by the sun and moon. These two bodies, one of which
lies always in the plane of the ecliptic and the other not far away,
pull upon the bulge in an attempt to straighten up the axis and
bring it into a position perpendicular to the plane of the ecliptic.
The combination of the pull of the sun and moon upon the earth's
bulge and the spinning of the earth upon its axis produces a force

31

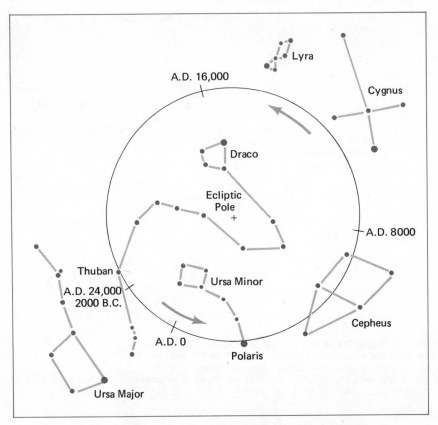

Figure 2.13 **The Motion of the North Celestial Pole**
As a result of precession, the north celestial pole traces out a circle in the sky.
At present the pole is near the star Polaris in Ursa Minor. At the time of
ancient Egypt, Thuban in the constellation of Draco was the "north star."

perpendicular to both these actions. This force is the precessional force, and when we describe the slow conical motion of the earth we say that the earth *precesses*. This means that the direction of the axis of the earth changes slowly. Indeed, the north celestial pole, which is simply the projection of the North Pole on the imaginary sphere of the sky, moves in a circle; the center of this circle is the pole of the ecliptic plane (Figure 2.13). The angular radius of the circle is about $23\frac{1}{2}°$, and the time needed for one complete precessional cycle is approximately 26,000 years. Thus the north celestial pole is moving slowly among the stars. At present it lies within a little less than a degree of a fairly bright star called the North Star, or Polaris, at the end of the handle of the

Little Dipper. In the years ahead the north celestial pole will move closer to Polaris. In the twenty-first century it will pass that star and begin to draw away from it. In the year A.D. 14,000, Vega, which is the brightest star in the northern sky, will be the pole star, although it will not be as close to the pole as Polaris is at present. Thousands of years ago, the star Alpha in the constellation of Draco, the dragon, was the pole star. During the age of exploration that started about the time of Columbus, Polaris was several degrees away from the pole. This rather large departure from the direction of true north occasionally caused confusion among mariners.

As the axis of the earth changes its position in the sky, the celestial equator, which is 90° away from that changing point, also shifts its position. As a result, the vernal and autumnal equinoxes —the points of intersection of the equator and ecliptic—move very slowly westward along the ecliptic. This means that the coordinates of the stars in any system based on the vernal equinox and the equator will change over time. For this reason, any good star catalog or star map will give the *date of the equinox* upon which it is based.

As a result of precession and the movement of the celestial equator among the stars, a considerable change has taken place over the centuries in what can be seen from a particular position on the earth's surface. Around 4000 B.C., the Southern Cross was visible in the latitude of New York; now, one must journey to the southern tip of Florida to see it in the sky.

Because the moon's orbit is slightly inclined with respect to the ecliptic, the moon is sometimes above or below this line. This changes the direction of the moon's attraction for the earth's bulge and produces a small *nutation* (literally, "nodding") of the pole over a period of 19 years, the approximate time it takes the line of intersection of the moon's orbit with the ecliptic to turn 360° with respect to the equinox. There is also a slight planetary precession induced by the tug of the other planets upon the earthly 'bulge that must be considered, but the major source of precession is still the pull of the sun and the moon upon the equatorial bulge.

## 2.9 TIME

For countless ages, man's only natural timekeeper was the sun, and the simplest instrument for recording the passage of time was the sundial. One of the earliest sundials was probably a vertical post with rocks appropriately distributed on the ground to mark the passage of the sun's shadow as it moved across the sky. Obviously, such an arrangement would be neither portable nor accurate enough for modern civilization, although many sundials are today considered both valuable antiques and works of art.

The sun as we see it, called the *apparent sun*, is a poor timekeeper. If a properly set up sundial is compared daily with an accurate clock, let us say at noon, the difference between the two will exceed 30 minutes at any given time during the course of a year. Assuming that the clock is accurate, the time it takes for the sun to make its apparent daily journey around the earth must be variable. This "journey" is caused by a composite of two motions: the earth's rotation once in about 24 hours and the earth's orbital motion about the sun. Measuring the rotation with respect to the stars rather than the sun eliminates the effect of revolution and shows the rotation to be so nearly uniform as to be incapable of producing the variation described above. However, the period of the earth's revolution is variable. Consequently, the sun's apparent motion must be variable also.

It might appear from what has been said that if the period of the earth's revolution were uniform (which it would be if the orbit of the earth were a perfect circle), then a sundial would keep uniform time. This is not so, because the axis of the earth's rotation is not perpendicular to the plane of its orbit. The apparent motion of the sun produced by the earth's rotation is measured

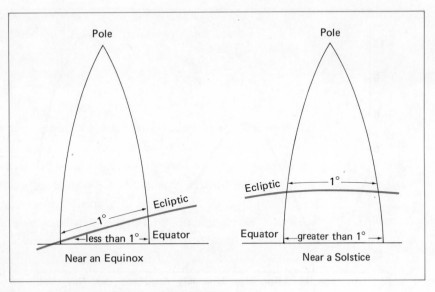

**Figure 2.14   The Relationship Between the Ecliptic and the Equator**
At or near an equinox, 1° of apparent solar motion along the ecliptic changes the
sun's right ascension (and hence solar time) by *less than* 1°. At or near a
solstice, 1° of apparent solar motion along the ecliptic changes the sun's
right ascension (and hence solar time) by *more than* 1°. Therefore, the apparent
position of the sun in the sky does not provide us with a method of keeping
uniform time throughout the year.

parallel to the equator, whereas the apparent motion produced by
the earth's revolution is measured along the ecliptic. (Figure 2.14
explains why uniform motion along the ecliptic, because of the
inclination of the earth's axis, is not equivalent to uniform motion
parallel to the celestial equator.)

Two circumstances, then, nonuniform revolution and the inclination of the earth's axis, combine to make the apparent sun a poor
timekeeper. In its place, astronomers use what is called the *mean*
sun, a fictitious body that keeps uniform time. The mean sun moves
in such a way that, in a year, it is ahead of the apparent sun as
often as it is behind it.

The circle on the celestial sphere from which we usually measure time is the *celestial meridian.* This is the circle passing through

**35**

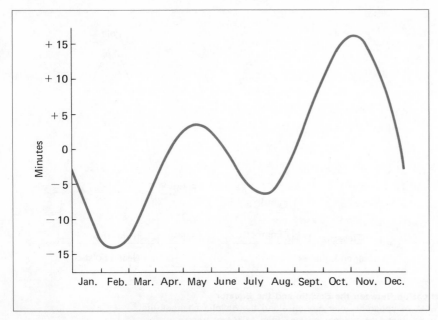

Figure 2.15 **The Equation of Time**
Only four times during the year are the mean and apparent suns together. The equation of time equals the apparent time minus the mean time. When the line on the graph is above zero, the apparent sun is fast. When the line is below zero, the apparent sun is slow.

the north and south celestial poles and the zenith. The celestial meridian is perpendicular to the *horizon*, the circle midway between the zenith and its opposite point, the *nadir*. The mean sun is on the observer's meridian at noon each day. An accurate clock set to indicate noon at this instant will then record noon each day as the mean sun *transits* (crosses) the half of the meridian that lies between the two poles and contains the zenith.

Such a clock is said to keep *mean time*. A sundial, on the other hand, keeps *apparent time*. The difference between the two is given by the *equation of time* (Figure 2.15). This variable difference (apparent time minus mean time) can range up to more than 15 minutes either way (depending on whether the apparent sun is fast or slow) and is zero only four times a year.

## 2.10 THE DETERMINATION OF TIME (STANDARD TIME)

Thus far, we have talked as if time were always derived from the observation of the sun, which is not true. Time is also determined from the stars. This kind of time is called *sidereal time* and is discussed in the next section. However, sidereal, or stellar, time can easily be converted to local mean time.

Standard time was first introduced in the latter half of the nineteenth century. With the advent of the railroads and the establishment of transportation networks throughout the United States, using local mean time to set schedules was very confusing. For example, Albany and Buffalo, New York, are approximately 280 miles apart. It takes the sun about 20 minutes to move these 280 miles westward. If local mean time were kept in each city, clocks in Buffalo would be about 20 minutes slower than those in Albany (and also those in New York City, which has nearly the same mean time as Albany). Figure 2.16 shows the various time zones in the United States.

To understand standard time, let us consider what happens as the sun moves westward for 24 hours. There are 360° in a complete circle around the earth. If there are 24 hours in the day, the sun must move 15° westward per hour, or 1° every four minutes. By international agreement, each 15° belt of longitude keeps the same standardized time. At sea this is called zone time, and the zone boundaries are straight lines—that is, circles around the surface of the earth.

These circles, or *meridians*, all of which pass through the north and south poles, delineate the terrestrial coordinate of longitude. Longitude is measured east and west of the meridian that passes through Greenwich, England (which has a longitude of 0°) to a meridian 180° away on the earth's surface. A zone 15° wide and extending from 7.5°E to 7.5°W of Greenwich keeps what is called *Greenwich mean time* (GMT). There are 24 such zones, all keeping

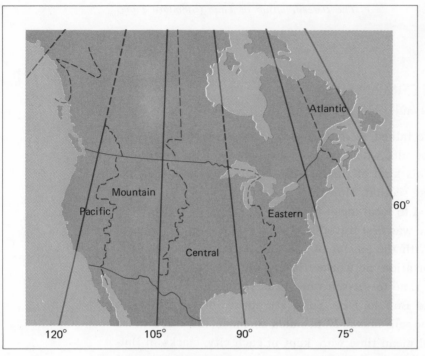

Figure 2.16 **Standard Time**
To avoid the inconvenience of a different local mean time for each meridian, the earth is divided into 24 more or less equal zones approximately 15° wide. At sea, the boundaries are meridians of longitude; on land, they are set by local option. The map shows the various time zones in the United States.

the mean time of their central meridians, whose longitudes are integral multiples of 15°, that is, 0°, 15°, 30°, 45°, and so forth. The zone boundaries are halfway between these multiples, at 7.5°, 22.5°, 37.5°, and so on.

The system described above is actually used at sea. On land, however, rather than being meridians, the zone boundaries are irregular. Usually, the time kept is that of a *standard* meridian whose longitude is a multiple of 15°. Hence the time in each zone differs from GMT by an integral number of hours. There are departures from this rule, however. For example, Newfoundland standard time is 3.5 hours slower than GMT.

A person journeying westward around the earth would have to set his watch back an hour each time he crossed a zone boundary. After he had completely circumnavigated the globe, he would have

38

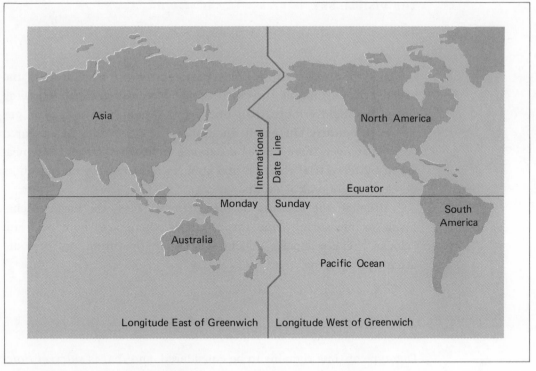

Figure 2.17 **The International Date Line**
A person circumnavigating the globe sets his watch back 24 hours if he is traveling westward and ahead 24 hours if he is traveling eastward. In order to be in step with time in his own country when he returns, he must also set his calendar ahead or back by one day. He does this as he crosses the International Date Line in the Pacific Ocean.

set his watch back 24 times. Having arrived back at his starting point, he would discover that his calendar did not agree with the local calendar. To prevent this anomaly, the *International Date Line* (Figure 2.17) was established in the Pacific Ocean. The International Date Line generally follows the 180th meridian, but in places it is diverted so that the date will be the same in all areas under the same government. A person crossing the International Date Line going from the American, or eastern, side to the Asiatic, or western, side compensates for 24 hours of setting his clock back by setting his calendar ahead. When it is Sunday on the American side of the dateline, it is Monday on the Asiatic side.

We have coordinated the measurement of time with the rotation of the earth, but due to various factors, the most important of which is the friction of the tides, particularly in the shallows of the oceans, the turning of the earth upon its axis is gradually slowing down, by slightly more than one one-thousandth of a second per day each century. In a few thousand years this fact has created somewhat over three hours' difference between time as it is measured by terrestrial rotation and time as measured by certain independent astronomical events.

Although the rotation of the earth has been the basic mechanism for measuring time for centuries, there are more accurate "clocks." These clocks depend for their operation on periodic atomic phenomena.

## 2.11 SIDEREAL TIME

One type of time remains to be considered, and that is *sidereal time*. Every observatory has among its equipment a sidereal clock that keeps time by the stars, or, more properly, by the position of the vernal equinox in the sky. The day begins as the vernal equinox crosses the upper meridian, and the interval until the next crossing is 24 sidereal hours. Sidereal time is the time used to calculate the positions of the stars in the sky at any instant.

We have already indicated that the apparent motion of the sun in the sky is produced by a composite of two motions: the rotation of the earth on its axis and its revolution about the sun. The earth rotates through 360° each day with respect to whatever clock, or timekeeper, we choose. For apparent time, our clock is the sun we see in the sky. Suppose now that the position of the sun at noon coincides with the vernal equinox. As the earth turns upon its axis and moves in its orbit, the sun will appear to move westward, because one rotation takes less time than one revolution. Suppose that one day passes with respect to the vernal equinox; that is, the earth rotates through 360° with respect to the vernal equinox. Because of the earth's revolution, the position of the sun will no longer coin-

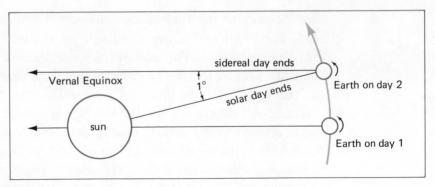

**Figure 2.18  Sidereal Time**
The sidereal day, which is the period of apparent rotation of the vernal equinox, is shorter by nearly 4 minutes than the solar day. This is because the revolution of the earth causes the apparent sun to move eastward along the ecliptic a little less than 1° per day. At the end of one sidereal day in the figure, the vernal equinox is again on the meridian, but the sun is still about 1° (4 minutes) east of the meridian. (For purposes of illustration, the 1° is exaggerated.)

cide with the vernal equinox. It will be about 1° east of the equinox, and about four minutes will elapse before it reaches the meridian (Figure 2.18). The first interval, from the time that the vernal equinox leaves the meridian until it returns to it, is called the *sidereal day*. The second interval, about four minutes longer, from the time that the sun is on the meridian until it returns to that point, is called the *apparent solar day*. Thus the solar day is about four minutes longer than the sidereal day, and there is one more sidereal day in a year than there are solar days.

## 2.12 THE CALENDAR

A *calendar* is a device that keeps track of the passage of time and divides it into short, more or less uniform, and convenient units. Historically, there have been three types of calendars: solar, lunar, and luni-solar. Our calendar is a *solar calendar*. It is based on the length of the solar, or *tropical, year*, which is the period between one passage of the sun through the vernal equinox and the next. Like the length of the day, the length of the year depends on the way it is measured. For example, suppose that we started to measure the year as the sun appeared to pass the vernal equinox, and, at the same time, we observed a fixed star exactly at the vernal equinox. A year later, when the sun had traversed the ecliptic and had returned to the vernal equinox, this point would no longer be coincident with the position of the star. This is because the vernal equinox moves westward along the ecliptic about 1° every 71 years. Thus the sun returns to the vernal equinox about 20 minutes earlier than it returns to the star. The tropical year, measured by the vernal equinox, lasts 365 days, 5 hours, 48 minutes, and 46.0 seconds. The *sidereal year*, the time that it takes the sun to move from a point coincident with the position of a particular star back to that same point, lasts a little longer—365 days, 6 hours, 9 minutes, and 9.5 seconds. This seems like a small difference, and perhaps it is; but it adds up over a period of time to a considerable amount. Using any period other than the tropical year to construct a calendar causes the dates of the seasons to shift.

A *lunar calendar* is based exclusively on the motion of the moon. The length of a lunar month (that is, the period of the phases of the moon, usually called a *lunation*), is nearly $29\frac{1}{2}$ days. The best example of a purely lunar calendar in use today is the Moslem calendar. This calendar has months 29 and 30 days long, which alternate through the year, so that the average length of a month is $29\frac{1}{2}$ days. Since $29\frac{1}{2}$ times 12 is 354, the Moslem year is approximately $11\frac{1}{4}$ days shorter than our $365\frac{1}{4}$-day year, and 33 Moslem years are equal to 32 of our years. Consequently, the dates of Moslem religious festivals change progressively in our calendar. Lunar calendars were more common in earlier days than they are today.

In a *luni-solar calendar*, such as the Hebrew religious calendar, the months stay in step with the lunations, but extra months are periodically inserted (intercalated) to keep reasonably in step with the solar, or tropical, year. In a period of 19 years, an ordinary solar calendar has 19 × 12, or 228, months. A lunar calendar, however, has almost exactly 235 months in 19 tropical years. To make up the difference, the Jewish calendar adds seven months during each 19-year cycle, intercalating them by a fixed pattern.

The ancient Roman calendar, which began with the legendary founding of the city of Rome, was a luni-solar calendar but was unsatisfactory because it did not have a fixed pattern for intercalating months. In 46 B.C. Julius Caesar abandoned it in favor of a solar calendar that averaged 365¼ days every four years. However, 365¼ days is still 11 minutes and 14 seconds longer than the true tropical year. This adds up to an extra day every 128 years. Thus between 46 B.C. and A.D. 1582, when the calendar was again changed, the date of the vernal equinox moved from March 25 to March 11.

The calendar reform of 1582 was the work of Pope Gregory XIII, called Gregory the Great. Gregory's goals were actually twofold. First, he wanted to bring the vernal equinox, which was used in determining the date of Easter, back to March 21, the date on which it had fallen in the year 325 (when the official ecclesiastical rule for determining the date of Easter was devised). He did this by removing 10 days from the calendar, making the day following October 4, 1582, October 15. To prevent a repetition of the problem, he decreed a revision in the number of leap years. The Julian calendar had achieved its 365¼-day year by having every fourth year include an extra day, so that the average length of four years was 365 days and 6 hours. Pope Gregory decreed that every fourth year should continue to be a leap year, with one notable exception. Only century years divisible by 400 would be leap years. Thus in both the Gregorian and the Julian calendars, the year A.D. 1600 was a leap year, and the difference between the two calendars remained 10 days. The years 1700, 1800, and 1900 were leap years in the Julian calendar but not in the Gregorian, making the difference between the two calendars 13 days. It will remain 13 days until the year 2100. Thus we now have a calendar that errs by one day in about 32 centuries.

# QUESTIONS*

(1) There are 88 modern constellations covering the entire sky. In what year, and by whom, were the boundaries of these constellations established? (Section 2.2)

(2) What method of designating the stars in a given constellation was suggested by Johann Bayer in 1603? (Section 2.3)

(3) What is the *Bonner Durchmusterung?* (Section 2.4)

(4) What kinds of objects are listed in the *New General Catalogue* and the *Index Catalogue?* (Section 2.4)

(5) Draw a schematic diagram of the celestial sphere. Indicate the locations of the north and south celestial poles, as well as the vernal and autumnal equinoxes. (Section 2.5)

(6) In what coordinate system are the terms *altitude* and *azimuth* used to indicate the location of an object in the sky? (Section 2.6)

(7) Define the terms *right ascension* and *declination.* (Section 2.6)

(8) For what purpose is the galactic coordinate system frequently used? (Section 2.6)

(9) Why are there seasons on the earth? (Section 2.7)

(10) What two factors are primarily responsible for the earth being hotter in summer than in winter? (Section 2.7)

(11) Why does a star catalog or star chart give the date of the equinox upon which it is based? (Section 2.8)

(12) Why is the sun a poor timekeeper? (Section 2.9)

(13) What is meant by "the equation of time"? (Section 2.9)

---

*The questions at the end of each chapter do *not* cover all the important concepts presented in the chapter. Rather, the purpose of these exercises is to give the reader a clear idea of the level of comprehension he should have acquired in the course of studying. The numbers located after each question indicate the chapter section that should be reviewed if a question cannot be readily answered.

(14) Why do we have time zones?         (Section 2.10)

(15) Which is longer: a sidereal day or a solar day? Why?

                                                (Section 2.11)

(16) Why do we have leap years?        (Section 2.12)

# CHAPTER 3
# THE NATURE AND USE OF LIGHT

Aside from what we have learned from examining various meteorites and a few hundred pounds of moon rocks, virtually everything we know about the universe beyond the earth comes to us from the study of light. Light from the stars and galaxies contains a tremendous wealth of information, and our understanding of the cosmos is in many respects limited only by the quality of our instruments for detecting light and our insight into its nature.

When we speak of radiation coming to us from outer space, we do not mean only visible light. The objects in the heavens also emit radio waves, ultraviolet and infrared radiation, x-rays, and gamma rays. Although these exotic radiations are invisible to our eyes, examining them by other means helps us to unlock the mysteries of our universe. All these forms of energy, including visible light, are called *electromagnetic radiation.* An understanding of the nature and properties of electromagnetic radiation is an obvious prerequisite to the study of astronomy.

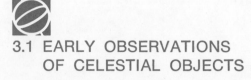

## 3.1 EARLY OBSERVATIONS
## OF CELESTIAL OBJECTS

Until the early seventeenth century, all observations of the positions of celestial objects were made using sighting rings and other graduated circles in combination with plumb bobs and other instruments for direct sighting without optical aid. Detailed observations of celestial objects were impossible without the magnifying or light-gathering power of lenses and mirrors. For this reason, the early history of astronomy is largely concerned with observations of the positions of objects rather than descriptions of their nature. Excellent work was done by many astronomers, and the motions of the moon, the sun, and the planets were surprisingly well understood. The length of the tropical year, for example, was determined with considerable accuracy before the introduction of the telescope. The Gregorian calendar, introduced some 28 years before Galileo's first use of the telescope, errs by only one day in 32 centuries. Ancient astronomers too did remarkable work in the astronomy of position. It is believed today, for example, that the mysterious Stonehenge in England was at least partly an astronomical observatory, and if some investigators are correct, an amazingly accurate and ingenious one. Moreover, there is some evidence that as many as 80 similar structures may have been built between 2000 B.C. and 1500 B.C. The astronomers who manned these observatories, however, worked only with the information transmitted by visible light, which is only a small part of the electromagnetic spectrum.

## 3.2 THE ELECTROMAGNETIC SPECTRUM

All the information we receive about celestial bodies is transmitted in the form of electromagnetic radiation. Most of the properties and reactions of such radiation are explained and understood most easily if they are discussed in terms of *waves*, although we do not mean to imply that such radiation consists purely of wave motion.

47

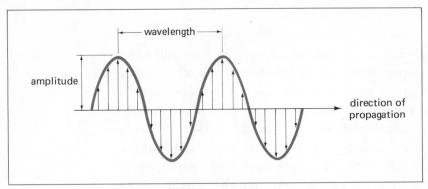

Figure 3.1 **Light as a Wave**
Light may be thought of as a wave phenomenon. The distance between successive peaks (or valleys) in the wave is the wavelength. The wavelength of electromagnetic radiation determines many aspects of its behavior.

Certain phenomena are more easily understood if we picture radiation as consisting of particles of energy called *photons*.

The basic property of any wave motion is its *wavelength*, that is, the distance between two successive, or adjacent, wave crests (Figure 3.1). If the velocity of the waves remains a constant, the wavelength is inversely proportional to the *frequency*, or the number of waves passing a fixed reference point in a unit of time. The inherent properties of any region of electromagnetic radiation are determined by its wavelength. Figure 3.2 shows the distribution, by wavelengths, of the various types of radiation. The longest waves are radio waves, which are usually measured in *meters* (one of which equals 39.37 inches) and kilometers (one of which is about five-eighths of a mile). Radio waves are produced on earth by special transmitters. Since they are long, their frequency is low, as is the energy of their photons.

Between about 7000 and 4000 *angstroms* (the symbol for which is Å) is the range of visible light. The angstrom is a convenient unit of measure when one wishes to discuss very small distances. One angstrom equals one hundred-millionth of a centimeter. Thus the wavelength of visible light ranges from about 0.000016 to 0.000028

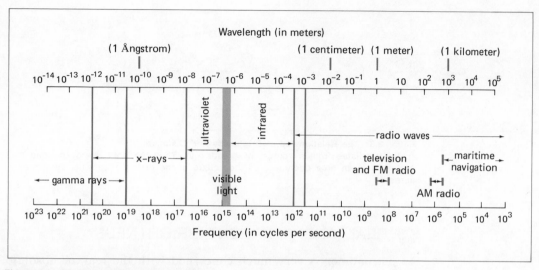

Figure 3.2 **The Electromagnetic Spectrum**
All electromagnetic radiation may be thought of as part of a continuous spectrum ranging from extremely short to extremely long wavelengths. Note the very narrow range of visible light.

inch. However, photographic films are available that are sensitive to radiations in the near infrared and ultraviolet portions of the electromagnetic spectrum, as well as to x-rays. Gamma rays have the shortest wavelength of all, and consequently have the highest frequency.

The speed with which electromagnetic radiation is transmitted is of particular interest to astronomers. It is not the same for all wavelengths, even in a particular medium transparent to the entire electromagnetic spectrum. A maximum speed is achieved in empty space. This is usually referred to as the speed of light in a vacuum and is designated by the letter $c$. Experiments shows that $c = 2.997925 \times 10^{10}$ cm/sec, or 186,282.1 mi/sec. The ratio of the speed of transmission of electromagnetic radiation in a vacuum to its speed in a transparent medium is called the *index of refraction* of the medium. Generally speaking, the higher the density of a medium, the higher its index of refraction.

49

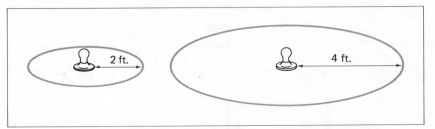

Figure 3.3  **The Relationship of Brightness and Distance**
As light radiates from a source, it spreads out. Thus the light energy passing through a given area decreases as the square of the distance from the source increases.

## 3.3 REAL AND APPARENT BRIGHTNESS

An astronomer is very much concerned with the real, or absolute, brightness of objects in the sky. It is impossible to tell how bright a star actually is merely by looking at it. It may truly be as bright as it seems, or it may be only a dim star that happens to be nearby. The *apparent brightness* of an object depends on its distance from the observer. To be more specific, the apparent brightness of a source of light varies inversely with the square of its distance. In other words, if a man observes a 100-watt light bulb at a distance of 100 feet, it will have a particular apparent brightness. If he then moves twice as far away, so that the distance between him and the bulb is 200 feet, the same light will appear only one-quarter as bright. On the other hand, if he moves toward the light until it is only 25 feet away, it will appear 16 times brighter than it did at a distance of 100 feet.

To see why this happens, imagine two identical light sources placed at the centers of two spherical shells, the first of which has twice the radius of the second. As Figure 3.3 shows, light emanating from the center of the larger sphere is spread over four times as much area as that emanating from the center of the smaller sphere. If the radius of the second shell were increased to three or four times that of the first shell, the same light would be spread over

9 or 16 times as much area. Hence stars with the same *intrinsic luminosity* have an apparent brightness inversely proportional to the square of their distance from us.

## 3.4 THE INTRODUCTION OF THE TELESCOPE

The time and place of the actual invention of the telescope is shrouded in mystery. We are fairly certain, however, that early in the seventeenth century a Dutch optician by the name of Lippershey discovered how to obtain an enlarged image of distant objects by placing properly adjusted lenses at the ends of a tube. The Italian physicist Galileo Galilei heard of this discovery and, from a description, made the first telescope used for astronomical observations. This was a major step in the history of astronomy, for it marked the beginning of the study of the nature of celestial objects as opposed to their position in the sky. Starting in 1610, Galileo made a number of startling discoveries. The best known of these are the four bright satellites of the planet Jupiter, called the *Galilean satellites.* He also investigated the nature of various features—particularly the mountains—on our own moon. He was unable to explain the protuberances on either side of the planet Saturn, which over a period of years seemed mysteriously to appear and then, just as mysteriously, disappear. In 1655, Huyghens found that this strange phenomenon was caused by the presence of a system of flat rings around the planet which were seen alternately on edge and tilted toward the observer.

As refinements were made in the telescope, the pace of astronomical discoveries accelerated. We shall take a look now at some of those refinements.

## 3.5 THE REFRACTING TELESCOPE

The telescope used by Galileo was a *refracting telescope*; that is, it used lenses which were circular pieces of glass with curved sur-

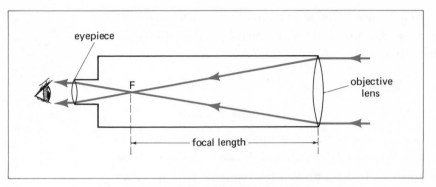

Figure 3.4 **The Simple Refracting Telescope**
The simplest form of refractor consists of two lenses, an objective lens and an eyepiece. The objective lens focuses the incoming light at the focal point, F. The eyepiece magnifies the image produced by the objective lens.

faces. When light passes obliquely through the boundary between media with different indexes of refraction, the rays are bent, or *refracted*. Figure 3.4 shows a cross section of the action of a *convex lens* (which is thicker in the middle than at the edges) on parallel rays of light from a distant object. Incoming rays from a far object are essentially parallel. After the refraction caused by entering and leaving the lens, which is called an *objective lens*, they converge at a focal point, F. The distance from the lens to the focal point is called the *focal length*. If a film is exposed at the focal point, the result will be a clear picture of the object.

If a lens with a short focal length, called an *eyepiece*, is placed so that its focal point coincides with that of the objective lens as in Figure 3.4, the emerging light rays will again be parallel, and we can view the image of the object directly. This is the principle on which the visual refracting telescope, or refractor, is based. The magnifying power of the telescope is equal to the focal length of the objective lens divided by the focal length of the eyepiece. Thus the combination of an objective lens with a large focal length and an eyepiece with a small focal length produces an instrument of considerable magnifying power.

Figure 3.5 **The 36-Inch Refractor**
This instrument, operated by the Lick Observatory, is a typical example of a
large refracting telescope. Lick Observatory photograph.

As techniques of manufacture improved, larger telescopes were
built, such as the 36-inch refractor in Figure 3.5. However, certain
problems developed in the larger instruments. We shall not discuss
all these problems, but we shall touch on one, called *chromatic aberration*. Although the light rays from a star are all bent by the objec-

53

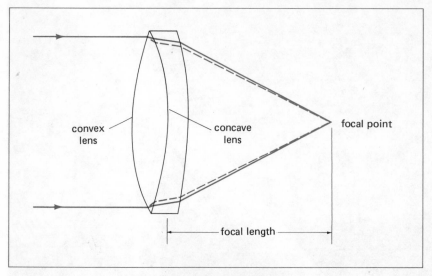

Figure 3.6 **An Achromatic Objective Lens**
The difference in chemical composition of the glass in two lenses corrects the chromatic aberration of the double convex lens alone.

tive lens, not all colors are bent by the same amount. Red light is bent the least, and blue and violet light the most. Consequently, instead of an image forming at a single focal point, images of different colors form at different focal points. This is known as chromatic aberration. It can be overcome, to a great extent, by combining two lenses, one doubly convex and the other *concave* (thinner at the middle than at the edges), as in Figure 3.6. The glasses in the two lenses have a different chemical formula, and the chromatic aberrations tend to cancel out. However, no refracting telescope is completely free of this defect.

Galileo's telescope was 2 or 3 inches in diameter. By 1860, telescopes 18 inches in diameter had been made. The largest refracting telescope (40 inches) was constructed in the 1890s, for the Yerkes Observatory in Williams Bay, Wisconsin. It is unlikely that a larger refractor will be constructed, since 40 inches seems to be the most efficient size for this type of telescope.

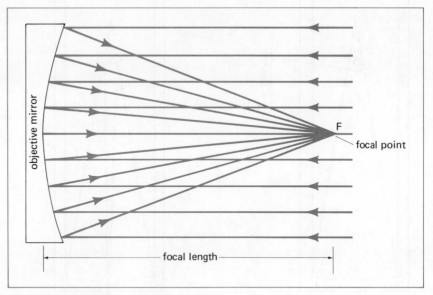

**Figure 3.7  The Image Formed by a Mirror**
A concave mirror ground in the shape of a parabola forms an image at point
F. There is no chromatic aberration. Astronomical mirrors are ground and
silvered (or aluminized) on the front surface. The light does not pass through the
glass, which therefore need not be of high optical quality.

## 3.6 THE REFLECTING TELESCOPE

Almost everyone is familiar with the fact that light is reflected by a
mirror. If the mirror is flat and made of a good grade of plate glass,
the image in the reflection is undistorted. If the mirror is curved,
the image is distorted. This fact is very useful in the construction of
astronomical telescopes.

All *reflecting telescopes* use a concave mirror, called an *objective mirror*, to focus light. The surface of this mirror may be a
spherical curve, but a better image is obtained if the curvature is
*parabolic*. A mirror ground in the form of a parabola forms a perfect, or nearly perfect, image of an object. If a source of light is infinitely far away, all the light travels in essentially parallel rays.
When these rays fall upon the parabolic surface parallel to the op-

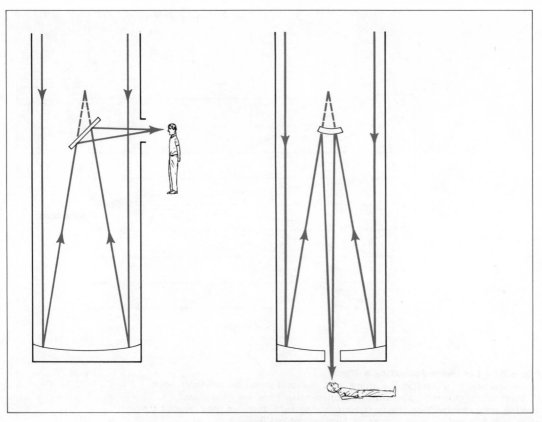

**Figure 3.8 Types of Reflecting Telescopes**
Since the image formed by a parabolic mirror, unlike that formed by the lens of the refractor, is in front of the objective, some device must be used to facilitate the observation without obstructing a significant portion of the incoming light rays.

tical axis of the telescope, an image is formed at a distance from the mirror called, like the distance from the lens to the focal point in a refracting telescope, the focal length. As Figure 3.7 shows, this point lies between the mirror and the source of light; hence the astronomer must reflect the image away from the light beam coming into the telescope.

The first reflecting telescope was made by Sir Isaac Newton. To reflect the image to the side of the telescope, he installed a mirror at an angle of 45° to the optical axis of the instrument and placed the eyepiece on the side. With a *Newtonian telescope,* one looks into the side of the instrument near the top in a direction perpendicular to the optical axis along which the telescope is directed (Figure

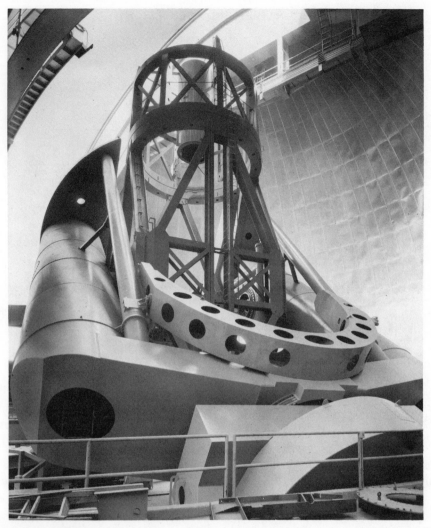

Figure 3.9 **The 200-inch Hale Telescope**
This instrument, the largest operational optical telescope in the world, is located on Mt. Palomar in southern California. The tubular capsule at the top is large enough for an astronomer to ride inside and work at the principal focus without using a diagonal mirror, as in the conventional Newtonian reflector. Photograph from the Hale Observatories.

3.8). Another type of reflecting telescope is the *Cassegrain telescope* (also shown in Figure 3.8), in which a slightly convex mirror is installed near the principal focus of the objective mirror and perpendicular to the optical axis. This mirror diminishes, or slows, the convergence of the light rays and reflects the light through a hole in

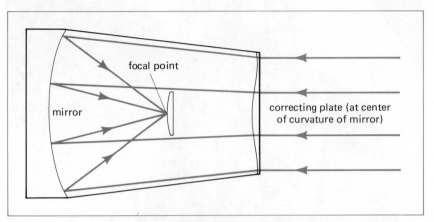

Figure 3.10 **The Schmidt Telescope**
In this instrument, the mirror is spherical, not parabolic, and the image is formed
in combination with a correcting plate. The Schmidt telescope gives sharp images
over a wide field and is especially useful in survey work.

the objective mirror. Since such an arrangement greatly increases
the effective focal length, a Cassegrain telescope has greater magnifying power than a Newtonian telescope. The largest reflecting telescope in operation today is the 200-inch Hale telescope on Mt. Palomar, near San Diego, California (Figure 3.9). There are several
instruments of this type with mirrors over 100 inches in diameter,
and the Soviet Union is working on a telescope with a 236-inch
mirror.

A special type of reflecting telescope, the *Schmidt telescope*,
overcomes at least one drawback of other instruments. Both reflecting and refracting telescopes give a clear picture of only a very small
region of the sky. As a result, when an astronomer attempts to make
large, wide-angle photographs, the definition deteriorates rather
rapidly with the distance from the center. The Schmidt telescope is
a wide-angle telescope with a spherical mirror rather than a parabolic one and a correcting plate in front of the mirror (Figure
3.10). It gives surprising detail to the edge of a rather wide field,
and has in many ways been a great boon to the astronomer. Figure
3.11 shows a Schmidt telescope in use at Mt. Palomar, California.

Figure 3.11 **The 48-inch Schmidt Telescope**
This very fine Schmidt telescope located at Mt. Palomar, California, was used in producing the *Palomar Sky Survey*. Photograph from the Hale Observatories.

## 3.7 WHAT A TELESCOPE DOES

A telescope does three things: It magnifies, it gathers light, and it resolves images. Each of these tasks is equally important in the efficiency of the instrument.

Perhaps the most obvious advantage of a telescope is that it magnifies faraway objects, making them seem nearby. As we noted briefly in Section 3.5, the *magnifying power* of a telescope is calculated by dividing the focal length of the objective lens (or of the primary mirror in the case of a reflector) by the focal length of the eyepiece. This statement can also be written as a formula:

$$\text{magnifying power} = \frac{\text{focal length of objective lens or primary mirror}}{\text{focal length of eyepiece}}$$

Thus if the objective lens or mirror of a telescope has a focal length of 50 inches and the eyepiece has a focal length of half an inch, the magnifying power of the instrument is 100.

Although magnification of the images of celestial objects is an important function of astronomical telescopes, large modern telescopes are built primarily to gather a large quantity of light from faint objects. The *light-gathering power* of a telescope is proportional to the area of its objective lens or the square of its *aperture* (the diameter of the objective lens). In other words, a reflecting telescope whose mirror is 10 inches in diameter gathers four times as much light as a reflecting telescope with a mirror 5 inches in diameter. To take another example, the iris of the human eye is about $\frac{1}{5}$ of an inch in diameter. The diameter of the 200-inch Hale telescope on Mt. Palomar is 1000 times larger; consequently, it has 1 million times the light-gathering power of the human eye. Only the huge telescopes in great observatories can reveal the faint (and frequently distant) objects so important to the study of astronomy today.

In addition to its ability to magnify an object and to gather light, a telescope also has *resolving power*, the ability to separate two close objects of moderate brightness. Viewed through a small telescope, the blurred images of two close stars may seem like a single bright object. Viewed through a larger telescope, however, this same bright object may appear to be two distinct and separate stars. In such a case, we would say that the larger telescope "resolved" the two stars. In general, the larger the telescope, the better its resolving power. The 200-inch Mt. Palomar telescope has twice the resolving power of the 100-inch Mt. Wilson telescope. Under ideal conditions, the Mt. Palomar telescope can resolve two stars separated by an angular distance of 0.023 second of arc, whereas the Mt. Wilson telescope can, at best, resolve stars separated by 0.046 second of arc.

The resolving power of a telescope decreases as the wavelength of the incoming radiation increases. The longer the wavelength, the harder it is to resolve images. Radio telescopes are particularly handicapped by this fact, and hence must be large to be efficient.

## 3.8 AUXILIARY TELESCOPE EQUIPMENT

The popular concept of an astronomer working long hours of the night, with his eye at the eyepiece of a telescope, is, alas, fallacious. Very little scientific observation is done visually today; instead, auxiliary equipment is used at the focal point of the telescope. Over time, various useful devices have been developed, notably the micrometer and the visual photometer. The greatest breakthrough, however, came with the introduction of photography to astronomy in the middle of the nineteenth century. A photograph has several advantages over a direct, visual observation. Properly taken, a photograph can be measured with a much higher degree of accuracy, and it automatically provides a permanent record of the observation.

*Photometry* is the branch of science concerned with the measurement of light. At first, photometry was done visually, by comparing a telescopic star image with a standard light source viewed simultaneously. Since it is easiest to match two light sources when they are equally bright, most visual photometers either dim the image of the star to match the standard source or vary the comparison source to match the star. This can be accomplished by simple mechanical means. Visual photometry is widely used and has the great advantage of being fast, easy, and economical.

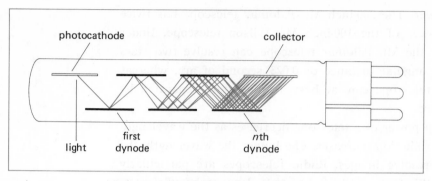

Figure 3.12 **The Photomultiplier Tube**
The photomultiplier tube is the basic component of all photoelectric photometers.
Light from the telescope is focused on the photocathode. When light strikes the
photocathode, electrons are emitted and are attracted to one dynode after another.
Each time an electron strikes a dynode, many more electrons are emitted. As a
result, the current at the last dynode is a million times greater than the current
at the first dynode.

In photographic photometry, the relative brightness of a star
is found by comparing either the density or the size of its photographic image to that of a standard sequence of stars of known
brightness. Photographic photometry is more accurate than visual
photometry, but it is also slower and requires more elaborate
equipment.

The apparent brightness of a celestial object, as measured by a
particular system of photometry, depends not only on the intensity
of the light emitted but on its color. A red light source, for example,
appears brighter to the eye than to the conventional photographic
plate. This fact eventually led to the development of a three-color
system of photometry, based on ultraviolet, photographic (blue),
and visual measurements obtained with filters and other equipment.
This *UBV system*, as it is called, has proved of great value to astronomers.

The most accurate photometry uses a *photoelectric cell* (Figure
3.12). This device, first used by astronomers during the 1920s and
greatly improved since, receives the starlight and produces a small
electric current. The electricity is amplified as much as a million

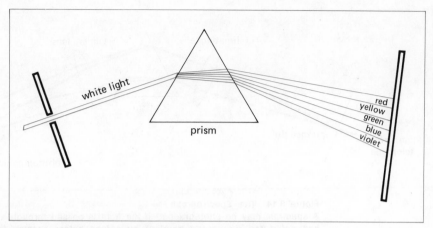

**Figure 3.13  The Action of a Prism in Forming a Spectrum**
Light rays traveling through a prism are refracted, or bent. Long wavelengths
(red light) are bent the least and short wavelengths (violet light) are bent the
most. Consequently, a beam of white light passing through a prism is broken up
into the colors of the rainbow, producing a spectrum.

times, until the signal is sufficiently strong to operate meters and
recording devices. A suitably calibrated photoelectric photometer
is the most accurate light-measuring device available to the astronomer.

By far the most productive auxiliary piece of equipment that
has been developed, however, is the *spectrograph*, and from the
science of spectroscopy has come a major portion, if not most, of
our knowledge of the nature of our universe.

Newton is said to have been the first to notice that when sunlight passes through a glass prism a rainbow of colors called a *spectrum* appears (Figure 3.13). No particular scientific use was made
of this phenomenon until some years after Newton's death, when
Joseph von Fraunhofer (1782–1826) discovered that if, before passing sunlight through a prism, he passed it through a fine slit to produce a narrow image, the spectrum would be crossed by dark lines
(Figure 3.14). These *Fraunhofer lines*, as they are called, indicate
that the white light coming from the sun is missing colors, or
wavelengths.

63

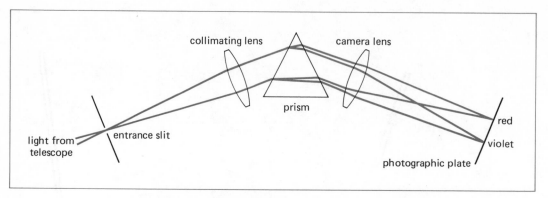

Figure 3.14 **The Spectrograph**
A spectrum may be photographed if the light is passed through a narrow slit
and collimated (i.e., made parallel) by a lens before passing through the prism.
A second lens then forms a series of sharp images on the plate or film. The
colors present in the light appear at different places in the spectrogram.

Further investigation showed that these dark lines could be duplicated in the laboratory by passing light from an artificial source
through various substances. Light from incandescent gases also produced lines that were bright against a dark background. Eventually,
Gustav Kirchhoff (1824–1887) formulated the following "three laws"
of spectral analysis:

(1) A luminous solid or liquid emits light of all
wavelengths and so produces a continuous
spectrum (Figure 3.15). (A gas under the high
pressure found in the sun and other stars may
also produce a continuous spectrum.)

(2) A luminous gas under low pressure produces
discrete bright lines (and sometimes a faint
continuous background).

(3) If light from a continuous source is passed
through a gas of lower temperature, the gas may
absorb certain wavelengths from the continuum
and produce dark lines in the spectrum.

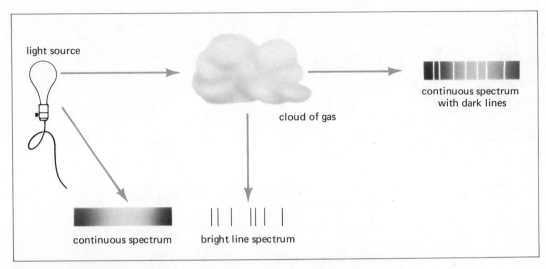

Figure 3.15 **Kirchhoff's Laws**
A hot object emits light at all wavelengths, producing a continuous spectrum.
After this light has been passed through a cloud of gas, certain wavelengths of
light are missing from the spectrum. The resultant dark lines are called
absorption lines. The spectrum produced by the illuminated gas, on the other
hand, consists of a series of bright lines called emission lines.

The bright lines produced by the luminous gas have the same wavelength (position in the spectrum) as the dark absorption lines in the spectrum of the continuous source whose light passed through the gas.

Kirchhoff's laws provide us with a very powerful tool for identifying the chemical composition of the atmospheres of the sun and other stars by their spectra. By comparing the lines produced by these stars with the lines of known elements, we can identify their chemical constituents.

A number of other properties of a light source may also be determined from spectral analysis. For example, changing the temperature of the source changes the character of the spectrum produced by a particular element. Therefore, if we observe the spectrum of a star, we can determine not only its chemical composition, but also its temperature.

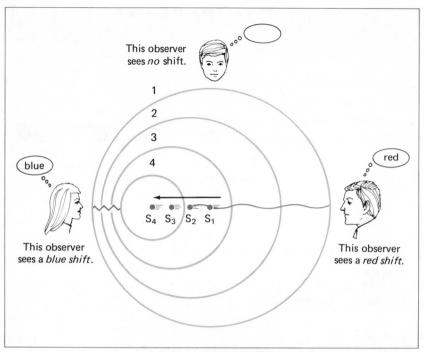

This observer sees *no* shift.

1
2
3
4

blue

This observer sees a *blue shift*.

red

This observer sees a *red shift*.

$S_4$ $S_3$ $S_2$ $S_1$

Figure 3.16 **The Doppler Effect**
When a source of light is approaching the observer, the waves are closer together than usual. The observed light therefore has a shorter wavelength and appears shifted toward the blue end of the spectrum. If a source of light is moving away from the observer, the waves are more spread out and the observed light has a longer wavelength and shifts toward the red end of the spectrum.

Another important piece of information that can be determined from the spectrum of a light source is the relative velocity of approach or recession of the source and the observer. This effect was first noticed in sound phenomena by Christian Doppler (1803–1853) and later applied to light. The *Doppler effect* causes spectral lines from an approaching light source to be shifted from their normal (at-rest) positions toward the violet end of the spectrum (Figure 3.16). If the source is moving away from the observer, the spectral lines are shifted toward the red end of the spectrum. The Doppler effect is therefore an extremely important tool in determining the velocities of stars and galaxies relative to the earth. An astronomer simply measures how much the lines in a particular spectrum are displaced from their normal positions. The greater the shift, the higher the speed of the light source.

Finally, when a beam of light passes through a strong magnetic field, the spectral lines are split into two or more parts. This is called the *Zeeman effect* and is useful in the study of sunspots and other light sources strongly affected by magnetic influences.

## 3.9 THE RADIO TELESCOPE

Until 1930, the only known electromagnetic radiation that was systematically observed coming from outer space was in or near the visible range of the spectrum. In 1932, Karl Jansky noticed that he periodically received radio signals from a source outside the earth. These were not signals in the ordinary sense of the term. To the radio astronomer, the word "signal" simply means electromagnetic radiation in the radio range. Even on earth, noise and radiation in this range exists apart from that generated by radio stations. We usually call most of this radiation "static." But the static, or signals, that Jansky received were actually coming from sources beyond the earth, indeed, for the most part, beyond the solar system. During World War II, British radar stations noticed radio noise from the sun, but this information was not released until 1946. Thanks to recent progress in radar and communications, radio astronomy has been one of the fastest growing branches of astronomy.

Radio telescopes are basically reflectors, but measured by standards of optical telescopes they are huge. Movable parabolic dishes larger than 200 feet in diameter are becoming relatively common (Figure 3.17). Huge paraboloids made up of fixed reflecting sections nearly 500 feet long have been constructed. The largest fixed radio telescope in the world, located at Arecibo, Puerto Rico, is 1000 feet in diameter. The device as a whole is not movable, but some directional adjustments can be achieved by moving the receptor.

We do not "see" objects with a radio telescope in the sense

Figure 3.17  **A 300-foot Radio Telescope**
This parabolic reflector is operated by the National Radio Astronomy
Observatory in Greenbank, West Virginia. Photograph courtesy of the National
Radio Astronomy Observatory.

that radiation strikes our eye or any other sensory human organ.
The function of a radio telescope is to gather the electromagnetic
radiation of a particular source and translate it into a voltage
whose magnitude and variation are recorded for study. What is
seen are the fluctuations of a needle or marking stylus as the telescope sweeps across the sky.

Radio telescopes have several advantages. They can be used
by night or day, since the reception of radio waves is not measurably affected by the sun. Radio waves also penetrate the gas and
dust clouds that obscure the central portion of the Milky Way and

various other portions of the galaxy as well. Objects beyond the range of optical devices can also be detected by radio equipment.

The spectrograph discussed in Section 3.8 and the ordinary household radio are actually closely related in principle. Both are instruments designed to arrange part of the electromagnetic spectrum in a linear fashion. It is then at the option of the operator to move along that range and pick out, or "tune in," a particular wavelength. If a radio receiver operated like the human eye, we would receive the entire range of radio energy to which the machine was sensitive. If this range happened to be the ordinary broadcast band, we would receive, simultaneously, the signals from all sources in that range. Listening to an individual station or program would be impossible. Radios, however, are equipped with a tuning device, and it is the ability of the particular piece of equipment to differentiate between signals that makes it possible for us to listen to the programs of our choice. Occasionally, when stations of equal power are close together on the tuning scale, confusion may result; generally, however, we are able, with a modern radio, to separate signals from the two sources successfully. The spectrograph does the same thing in the visible and photographic ranges of the spectrum. The sensitivity of a spectrograph depends on the quantity of light from the optical system and the dispersion, or spreading power, of the spectrograph itself. In a sense, making a stellar or solar spectrogram is like moving the tuning dial on a radio. Although the study of other parts of the electromagnetic spectrum also has its uses, we shall be primarily concerned here with contrasting and comparing radiation in the range of visible light and the radio range.

We discussed the problem of resolving power to some extent in Section 3.7. As we noted, resolving power is the ability of a particular detecting system to separate radiation from sources close together in the sky. Large objective lenses have a high resolving power. The human eye, binoculars, and small telescopes have relatively poor resolving power. The mathematical formula for resolving power depends not only on the aperture of the equipment used but on the wavelength of the radiation being analyzed. Other things being equal, an increase in the wavelength of the radiation produces a proportionate decrease in resolving power. In other words, the resolving power of any particular telescope is theoretically greater for radiations in the ultraviolet range than for infrared signals.

The radio portion of the spectrum consists of waves whose length is measured in centimeters, meters, and even kilometers. As a result, radio telescopes must be large to have even modest resolving power. The largest optical telescope in use today is 200 inches, or approximately 5 meters, in diameter. The largest radio telescope is 1000 feet, or approximately 300 meters, in diameter. However, its resolving power does not approach that of the 200-inch optical telescope. (Indeed, for a radio telescope to have the same resolving power as a 200-inch optical telescope, its diameter would have to be about 10,000 kilometers, or 6,000 miles, nearly three-fourths the diameter of the earth.) This makes the identification of a celestial radio source with a known optical source a difficult problem.

## 3.10 INTERFEROMETRY

One way to get around the low resolving power of radio telescopes (at least for certain purposes) is through *interferometry*. Two antennas some distance apart along a baseline are used to record the signal from the same radio source. If the source is not on the perpendicular bisector of the baseline, the two signals will not be received exactly in phase. That is, they will be received by one antenna slightly before they are received by the other antenna. Since the extent to which the signals are out of phase (or interfere with each other) depends on the angle an imaginary line in the direction of the source forms with the baseline and the length of the baseline, the direction of the source can be determined with great accuracy.

The first astronomical use of interferometry was at the Mt. Wilson Observatory, where the diameters of giant stars in the earth's neighborhood were measured by adjusting the separation of the receiving mirrors on a crossbar attached to a 100-inch telescope. Radio telescopes mounted on railroad cars and separated by a known amount were used by radio astronomers in the same way. In recent years, telescopes as far apart as Sweden and the United States have been used as the base for interferometry. For at least some purposes, this separation has the same effect as having one telescope equal in diameter to the distance between the two instruments.

# QUESTIONS

(1) Name five types of radiation in the electromagnetic spectrum. (Section 3.2)

(2) In what way is the apparent brightness of a source of light related to the distance between the source and an observer?
(Section 3.3)

(3) Draw a diagram showing how a refracting telescope works.
(Section 3.5)

(4) Draw a diagram showing how a Newtonian reflecting telescope works. (Section 3.6)

(5) Describe three functions of a telescope. Why is each of these functions important? (Section 3.7)

(6) What is a spectrograph? (Section 3.8)

(7) What are Kirchhoff's laws? Why are they important?
(Section 3.8)

(8) What is the Doppler effect? What sort of information does it give us? (Section 3.8)

(9) Describe three advantages of radio telescopes. (Section 3.9)

(10) One of the several drawbacks of radio telescopes is their low resolving power. How is interferometry used to overcome this difficulty? (Section 3.10)

# CHAPTER 4

# THE WORKINGS OF THE SOLAR SYSTEM

Ancient observatories are found among the relics of many civilizations, each of which produced its own cosmology of creation, and we can trace the steady development and astronomical ideas and understanding. One of the first problems to intrigue astronomers was that of the motions of the sun, the moon, and the planets, the wandering objects in the sky. In this chapter we shall consider the genesis and development of the most popular theory of the Middle Ages regarding these heavenly bodies and its overthrow by what is now the accepted explanation of the solar system.

We shall study the motion of the system of planets surrounding the sun. This system is called the *solar system* because of the dominance of the central star, the sun—a dominance that has greatly affected the development of astronomical theory.

Throughout antiquity, various ideas were propounded to explain the observed behavior of the planets. The dominant theory of Roman times was the *Ptolemaic theory.* This world view was not successfully challenged until late in the fifteenth century, and it was not until about 250 years later that observational evidence decisively settled the conflict.

## 4.1 THE PTOLEMAIC SYSTEM

Claudius Ptolemy, in the second century A.D., described a system of planetary motions in *The Almagest*. This system, like most ancient theories of the movement of heavenly bodies, was geocentric, regarding the earth as the center of the universe. It assumed that all the planets moved with uniform speed in circular orbits about a motionless earth. These circular orbits were called *deferents*. The "planets," in order outward from the earth, included the moon, Mercury, Venus, the sun, Mars, Jupiter, and Saturn.

However, the observed behavior of the planets by no means coincided with this simple system. The planets seemed to accelerate and decelerate, sometimes even reversing their generally eastward motion among the stars. A simple system of deferents could not explain this variable motion in the sky. To overcome the problem, *epicycles* were introduced (Figure 4.1). The planets were said to move in smaller circles about a point which itself moved uniformly along the deferent. By choosing an appropriate ratio of the size of the deferent to the size of the epicycle and the ratio of the periods of revolution as well, the planet's observed motion could be approximated. However, as observations improved, the difference between the predicted and observed planetary positions became intolerable. Epicycles proliferated. Some were eccentrically placed on their preceding counterparts; others were inclined to the principal plane of the system to explain observed departures from the ecliptic by the planets. The planets Mercury and Venus are never very far from the sun; their *elongations* (the angle between them and the sun) is never very large. To prevent large elongations of Mercury and Venus, the centers of their first epicycles had to remain on a line joining the earth and the sun.

Although the Ptolemaic system worked well enough for the early astronomers, whose observations were made with the naked eye, it was unwieldly. Nevertheless, it remained the prevailing astronomical doctrine for well over a thousand years.

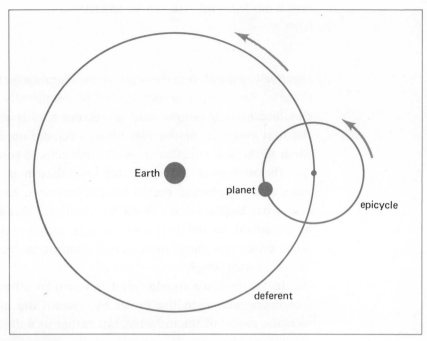

Figure 4.1 **The Ptolemaic System**
Each planet moves along an epicycle, which in turn moves along a deferent.
The earth is at or near the center of the deferent. As the system was refined,
further epicycles had to be added.

## 4.2 EPICYCLES AND THE PATHS OF PLANETS

To understand the problem faced by ancient observers, let us consider how the visible planets move in the sky. We shall assume for
the moment, as did the ancients, that the stars form a fixed background against which the planets move.

The moon, closest to the earth and moving most rapidly, presents the least problem. Its elliptical path gives rise to some variable
motion, but this can be explained by a few epicycles. Moreover, the
inclination of these epicycles can be made to account for the observation that the moon at times is over 5° off the ecliptic. The sun,
which is always on the ecliptic, needs no inclined epicycles, and its
variable speed along the ecliptic can be reasonably well explained
by a few epicycles.

Mercury and Venus have the curious property of being seen
only near the sun. Their maximum elongations are 28° and 47°, re-

spectively, and Mercury's orbit is measurably inclined (7°) to the ecliptic. However, a proper choice of empirically calculated epicycles, eccentric locations, and inclinations, plus the provision that the first epicycles of the two planets remain on a line joining the earth to the sun, explains these observations to some extent.

The motion of Mars, Jupiter, and Saturn is in most respects more easily explained, except for the fact that Mars' actual orbit is somewhat inclined and somewhat elliptical. This last property can be explained, in the Ptolemaic system, only by the addition of so many cycles that the system seems in imminent danger of collapsing under its own weight.

In fairness, we should point out that in all probability the astronomers who used the Ptolemaic system did not regard it as a working model of the universe, but rather as a device for predicting planetary positions. The choice of the dimensions of the various parts was based on experiments and observation. Hopefully, the calculations, when carried forward, would predict the future planetary positions.

The Ptolemaic system accomplished its task with reasonable success for 1200 years. The fact that it took nearly 200 years for the more successful Newtonian system to replace it attests to its service to astronomers through the centuries. Eventually, however, revisions to take new observations into account were not enough. The time for a basic change was at hand.

## 4.3 NICOLAUS COPERNICUS (1473–1543)

The year 1543 was a memorable one for astronomers, for it marked the publication of a book entitled *On the Revolutions of the Celestial Bodies*, written by the Polish scientist Nicolaus Copernicus (Figure 4.2). In it, Copernicus showed that all the motions of celestial bodies could be explained more rationally if the sun, rather than the earth, was the center of the solar system. He went on to propose

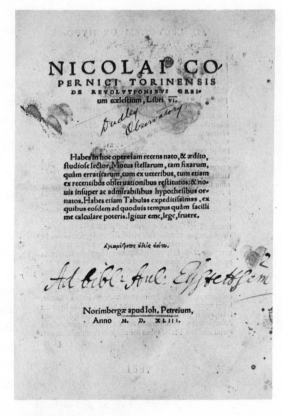

Figure 4.2 **The Title Page of Copernicus' *De Revolutionibus Orbium Coelestium***
(1st ed., Nuremberg, 1543)
Dudley Observatory Photograph.

that the geocentric orientation of the Ptolemaic system be discarded
and a heliocentric interpretation attempted. The idea that the earth
was not the center of the universe was not entirely new, but it went
against the religious doctrines of the day and began a controversy
that continued for almost two centuries.

Copernicus made no effort to dispense with the systems of epicycles or with the notion of circular orbits. On the contrary, he retained them as a necessary explanation of the irregularities and directional changes in the observed motion of planetary objects. He
offered no proof that his system was more likely to be true than its
predecessors. His book was simply an appeal to reason. To him, it
appeared more reasonable for the earth to turn about an internal
axis than for the Universe to revolve around the earth. His ideas

were not accepted universally at once; it was years after his death before his theories were vindicated. Indeed, at first, the evidence seemed to indicate that Copernicus was wrong.

## 4.4 TYCHO BRAHE (1546–1601)

In the second half of the sixteenth century, there lived in the southern part of what is now Sweden one of the best astronomical observers of all time, Tycho Brahe. Although his observatory was the finest in the world at that time, Tycho died before the application of the telescope to astronomy. Thus he could know nothing of the nature of the objects he saw. However, he made careful observations of the various planets and their positions, using large sighting circles, quadrants, and other instruments. Night after night, he measured the positions of the planets with reference to the fixed stars. He was particularly interested in the planet Mars.

Tycho set about to test the Copernican theory, reasoning that if the earth were moving about the sun, there should be a detectable *parallax*, or shift in the positions of the stars. In other words, if the earth revolved about the sun, the nearby stars should appear to move back and fourth slightly over the course of a year. However, Tycho had no concept of the actual distances of stellar objects. He believed that if the stars were no more than 7,000 times as far away from the earth as the earth was from the sun, he could observe their parallax with his instruments. This was true. But the nearest star was not 7,000 times as far as the sun; it was about 270,000 times as far. Such distances were way beyond the reach of Tycho's instruments. His logic was correct, but his premise and conclusions were wrong. Since no parallax was discernible, Tycho concluded that the earth was the center of the solar system and that the sun and moon revolved around the earth. However, he suggested that objects other than the sun and the moon be regarded as moving about the sun. This view of the solar system was called the *Tychonic*

Figure 4.3 **The Title Page of Brahe's** *Astronomicae Instauratae Progymnasmata*
(Frankfurt, 1610)
Dudley Observatory Photograph.

*system*. The differences between it and the Copernican system could not be detected until telescopes and other optical aids had been developed. Figure 4.3 shows the title page of Tycho's work.

## 4.5  JOHANNES KEPLER (1571–1630)

Johannes Kepler, a German, served as Tycho's assistant for about a year, and when Tycho died Kepler came into possession of his notes and the records of his observations. Few figures in the history of science are more controversial than Kepler. Some regard him as a

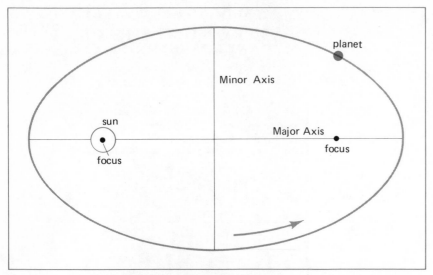

Figure 4.4 **Kepler's First Law**
Kepler's first law of planetary motion stated that the orbits of the planets were
ellipses, with the sun at one focus, and not circles, as astronomers had
formerly believed.

genius of the first order; others see him as a charlatan. Certainly he
at least pretended to believe in the pseudo-science of astrology, and
he would cast horoscopes for a fee. His importance to astronomy,
however, lay in his formulation of the following "three laws" of
planetary motion:

(1) The orbit of each planet is an ellipse, with
the sun at one focus. (Figure 4.4)

(2) Each planet revolves so that a line connecting
the planet earth with the sun sweeps out equal
areas in equal time. (Figure 4.5)

(3) The squares of the periods of any two planets
have the same ratio as the cubes of their
average distance from the sun. (Figure 4.6)

The first two laws were published in 1609 and the third nine
years later. The first law disposed of the notion of uniform circular

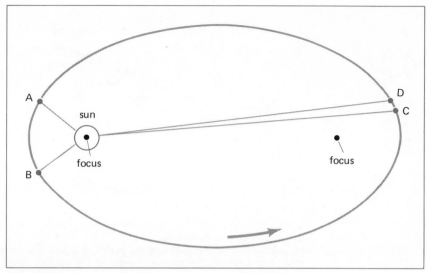

**Figure 4.5 Kepler's Second Law**
This law stated that the speed of the planets as they revolved about the sun was such that a line connecting each planet and the sun swept out equal areas in equal intervals of time. The two triangles in the figure, although they have a different shape, have the same area. Thus it takes the planet as long to get from A to B as it does for it to get from C to D.

motion. It did not necessarily assert that the earth was a planet, leaving this to be established later. The second law helped to explain why the planets, and particularly the moon, did not move at a uniform rate. The third law provided astronomers with a simple method of determining the distance of the planets from the sun, once it was established that the earth was a planet.

Of Kepler's three laws, the third is perhaps the most difficult to understand. It can be stated fairly simply in mathematical terms, but first we must make note of some definitions. The average distance from the earth to the sun is 92,960,000 miles. This distance is equal to one *astronomical unit*, or 1 AU. As we shall see, the AU is a very convenient yardstick in discussing the solar system. The earth's orbital period, the time that it takes the earth to move from a fixed point in space back to that same point, is 365 days, 6 hours,

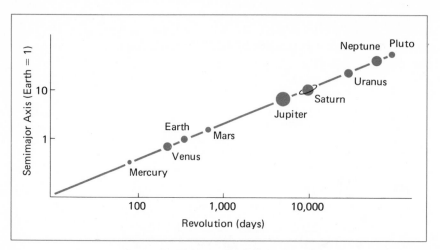

Figure 4.6 **Kepler's Third Law**
This graph shows a plot of the orbital period of a planet about the sun
against the length of the semimajor axis of the planet's orbit. The fact that the
data from all the planets lie along a straight line tells us that there is a simple
relationship between these two quantities. Kepler's third law predicts the exact
orientation of this line.

9 minutes, and $9\frac{1}{2}$ seconds. This is called a *sidereal year*. Just as the
AU is a convenient unit of distance, so the sidereal year is a convenient unit of time.

Kepler's third law may be expressed by the formula

$$d^3 = p^2$$

where $d$ is the average distance from the sun to a planet measured
in AU's and $p$ is the orbital period of the planet in sidereal years.
For example, the orbital period of Jupiter is 11.86 sidereal years. It
takes Jupiter this long to go all the way around the sun and return
to its starting point. Thus, for Jupiter, $p = 11.86$, and $p^2 = 140.7$. Kepler's third law tells us that $d^3 = 140.7$; therefore, $d = 5.20$. Consequently, the average distance from the sun to Jupiter is 5.20 AU's,
or about 483 million miles. Since the orbital period of any planet
can be obtained from direct observation, this method can be used
to determine the relative size of the planets' orbits. Unfortunately,
no one in Kepler's time knew the distance from the earth to the

sun. Indeed, for many years following Kepler, no sound information was available regarding this fundamental unit of solar system. Even today, the definition of the AU is still being refined.

## 4.6 GALILEO GALILEI (1564–1642)

Galileo, the great Italian physicist, was interested in all branches of science. He was an enthusiastic proponent of the experimental method. Among his great contributions to the science of astronomy are the adaptation of the telescope to celestial observations, experiments in the movement of bodies near the surface of the earth, and observations of the movement of the celestial bodies. With his small telescope, he discovered, in January 1610, the four bright satellites of Jupiter known today as the *Galilean satellites*. This discovery lent support to Copernicus' ideas, for here, in miniature, was a planetary system revolving about one of the members of our solar system. Galileo also observed the planet Venus and found that its phases, which were clearly visible in his telescope, contradicted the hypothesis of the Ptolemaic system and confirmed Copernicus' statement that Venus revolved around the sun. (Through his telescope, Galileo was able to observe that Venus passed through four phases in its journey around the sun, full, quarter, crescent, and gibbous. The full and gibbous phases are impossible in the Ptolemaic system.) Finally, Galileo's investigations of the motions of bodies and particles in general led him to the conclusion that they moved uniformly in a straight line and deviated from their straight-line motion only if a *force* was exerted on them. This conclusion was later to be the basis of Isaac Newton's first new law of motion.

## 4.7 ISAAC NEWTON (1642–1727)

The greatest of all the astronomers and scientists concerned with the Copernican revolution, and the last that we shall consider, is Sir

Isaac Newton. Born the year that Galileo died, Newton coordinated the investigations of others. He contributed enormously to the progress of science in general and mathematical physics in particular. It is to him that we owe the law of gravitation, a theory that dominated physics for 200 years.

Newton appeared on the scene at a time when science was ripe for basic changes in many areas. Geometry, the chief mathematical tool of the Ptolemaic system, provided an inadequate explanation of findings in astronomy and physics. New tools were needed, and Newton participated in their invention. Using formulas based on what we now call calculus, he constructed a model of the solar system that was not dependent primarily on geometry. By consolidating the work of his predecessors and strengthening it with his own contributions, he laid the foundations for further progress. His most famous contributions are the three laws of motion described below and the law of universal gravitation.

## 4.8 THE THREE LAWS OF MOTION

Late in the seventeenth century one of the most important books ever to appear in science was published, Sir Isaac Newton's *Principia Mathematica*. It contained the following three laws:

(1) Every body continues in a state of rest or of uniform motion in a straight line unless that state is changed by the action of a force upon the body.

The foundations of this law were laid by Galileo. Until his time, most people believed that in order for an object to move, a force had to be applied to it; if there were no force, the body would come to rest. Galileo stated that uniform motion in a straight line requires no force, and once a body starts moving uniformly in a straight line, it will continue to do so for an infinite time. Since there is no way of determining the absolute position of an object in space, there is no way to tell whether a particular body is at rest or moving unless a reference point is chosen. The body may then be said to be at rest or moving with respect to that reference point.

> (2) The acceleration of a body is directly proportional to the net force acting on the body and inversely proportional to its mass. Furthermore, the acceleration takes place in the direction of the straight line in which the force acts.

This law is stated algebraically as

$$F = ma$$

where $F$ is the force, $m$ is the mass of the body, and $a$ is the acceleration of the body acted on. This law may be used to determine the mass of a body when $F$ and $a$ are known.

> (3) For every action, there is always an equal and opposite reaction.

Thus the earth attracts the moon and the moon attracts the earth by an exactly equal force. However, the effects of equal forces are not always the same. Since the mass of the moon is smaller than the mass of the earth and the forces with which the two bodies attract each other are equal, the acceleration of the moon is greater than the acceleration of the earth.

## 4.9 THE LAW OF GRAVITATION

Newton's mathematical tools and his three laws of motion eventually led him to the law of universal gravitation, which is

> Every particle in the universe attracts every other particle with a force that varies directly with the product of their masses and inversely with the square of the distance between them. (Figure 4.7)

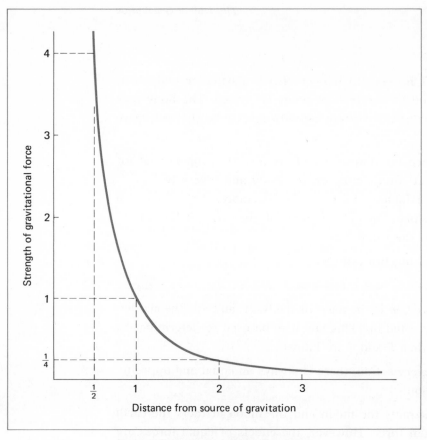

Figure 4.7  **Newton's Law of Gravitation**
The gravitational force of a body diminishes with our distance from it. If we move twice as far away, the force is only one quarter as great.

This law may be expressed mathematically as

$$F = G\,\frac{m_1 m_2}{d^2}$$

where $G$ is the constant of gravitation and has a particular value determined by experimentation in a laboratory, $m_1$ and $m_2$ are the respective masses of the two bodies, and $d$ is the distance separating them.

Every physical object that has mass also has a *gravitational field*, or area of influence. If the mass of the object is doubled, the strength of the field also doubles. The strength of a gravitational field also depends on the distance from the object generating the

86

field. An observer close to a massive object experiences a certain gravitational force. If he moves twice as far away from the object, the strength of the gravitational force will be only one quarter as great.

Newtonian physics, based on the laws of motion and gravitation described above, dominated research in many areas of astronomy from the middle of the eighteenth through the early part of the twentieth century. The branch of astronomy known as *celestial mechanics* developed techniques that enabled astronomers to predict with precision the observed motions of bodies both in and without the solar system. The two centuries that elapsed after the work of Newton was well established saw almost uninterrupted progress in this area. Using the law of gravitation and the mathematical methods he developed, Newton was able to derive all Kepler's laws. He explained Kepler's discoveries in terms of an attractive gravitational force that all material bodies (not just planets) must possess. Furthermore, he explained and predicted a number of phenomena that had not previously been understood or known. For example, an ellipse is one member of a family of curves known as *conic sections*. Newton proved that the orbit of an object about the sun could be any one of the conic sectons: a circle, an ellipse, a parabola, or a hyperbola. The precise orbit of a celestial object about the sun would depend very specifically on how much energy the object possessed. Objects with little energy (compared to the energy in the gravitational field of the sun), such as planets, must orbit the sun in circles or ellipses. Objects with a great deal of energy, such as comets, can have orbits that are parabolic or hyperbolic.

## 4.10 THE SOLAR SYSTEM

We come now to the task of describing the group of objects that we consider to be members of the solar system. Our descriptions will be in terms of Newtonian physics, and we shall assume that the Newtonian explanation of the physical universe holds throughout its entirety.

At the center of the solar system is the *sun*. The sun is by far the most massive member of the system and dominates not only the planets but a region of space much larger than that occupied by the solar system itself. The mass of the sun constitutes over 99.8 percent the total mass of the solar system. Thus all the other bodies it contains combined account for less than 0.2 percent of its mass. As a result, in most mathematical calculations of the orbits of the planets, we can neglect the mass of a planet compared to the mass of the sun. This frequently simplifies the calculations greatly. The study of planetary motion is also simplified by the isolation of the solar system. The nearest star to the sun is 26 million million miles away. This isolation means that the gravitational attraction of any body other than the sun is negligible and can be ignored. The effects of the planets on one another can be neglected in all first approximations of planetary motions, but they must be considered in highly precise calculations.

## 4.11 THE PLANETS AND THEIR ORBITS

Certain statements can be made about the planets and the orbits they follow around the sun. The planets, in distinction to the sun, are cold bodies that shine only by reflected sunlight. For the most part, they depend entirely upon the sun for heat and light. Without the light of the sun, the planets would be invisible. Their temperatures would drop to that of empty space, and life and movement upon them would be quite impossible. It is only because of the heat of the sun that life has been able to develop on the earth's surface.

Modern astronomers classify the sun as a star and the moon as a satellite. The earth has been added to the original list of

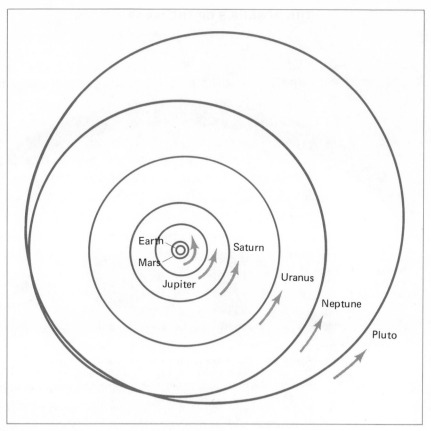

Figure 4.8 **The Orbits of the Planets**
This scale drawing of the solar system shows the great range in the size of the
orbits of the planets. (The orbits of Venus and Mercury, which are within the
orbit of Earth, are not shown.) All the planets, with the exception of Pluto, have
their orbits in very nearly the same plane. Pluto's orbit is inclined by some 17° to
the general plane of the solar system.

planets, as have Uranus, Neptune, and Pluto, which were discovered,
respectively, in the eighteenth, nineteenth, and twentieth centuries.
Thus there are now nine known planets. It seems unlikely that
more will be discovered, unless there are small planets revolving
in orbits well beyond the outermost known planet, Pluto. Two
planets, Mercury and Venus, are closer to the sun than the earth,
and six are further away. The planets differ widely with respect
to size, physical constitution, orbits, and other characteristics, but
they do have certain similarities.

All the major planets revolve about the sun in elliptical, nearly

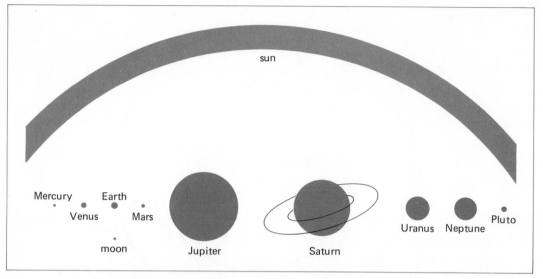

Figure 4.9 **The Relative Sizes of the Sun and Planets**
This schematic diagram shows the relative size of the sun and its planets.
Mercury, about the size of our moon, is the smallest. Jupiter, with a diameter
eleven times that of the earth, is the largest.

circular, orbits (Figure 4.8). Pluto has the largest orbit, about
40 AU's, and Mercury the smallest (0.4 AU). The orbits of all the
planets lie in or near the plane of the ecliptic. Their inclinations,
except for the planet Pluto, are moderate. Furthermore, the planets
all revolve in the same direction, counterclockwise as viewed from
the north celestial pole. Most of the planets rotate in a counterclockwise direction as well, and the majority of the 32 known
satellites in the solar system revolve about their primaries in this
direction. These uniformities and similarities in the orbits and
movements of nearby celestial bodies are fortunate, for they simplify
the task of explaining the nature and origin of the solar system.

Mercury, Venus, Earth, and Mars are generally referred to as
the four *inner planets*. They range in diameter from about 3000 to
8000 miles. They are similar in some ways and different in others.
Mercury, the smallest, resembles our moon and is believed to have

no atmosphere. Venus, about the size of the earth, has a heavy atmosphere that efficiently reflects the light of the sun. As a result, Venus periodically becomes a very bright object in the sky. Mars is intermediate between the moon and the earth in atmospheric and surface characteristics and has a reddish tinge. Mars, which is farther from the sun than Earth, does not appear as bright as Venus.

Four planets, Jupiter, Saturn, Uranus, and Neptune, are classified as *major planets*. They are, as Figure 4.9 shows, relatively large, ranging in diameter from approximately 28,000 to 88,000 miles. The largest is Jupiter, over 11 times the diameter of Earth. Saturn is slightly smaller than Jupiter. Uranus and Neptune, which lie still further from the sun, were discovered after the invention of the telescope. Pluto, a small planet and the most distant of all, spends most of its time beyond the orbit of Neptune. Because Pluto's orbit is rather eccentric, it does at times come closer to the sun than Neptune. All the major planets are somewhat similar physically, with heavy atmospheres, rapid rates of rotation, and more oblate shapes than the smaller planets.

## 4.12 ASPECTS AND PHASES OF THE PLANETS

For our purposes, the planets can also be classified in terms of their *aspects*, their situation with respect to one another and other objects in the solar system. Omitting the earth momentarily from consideration, we may speak of the planets as being either inferior or superior. An *inferior planet* is a planet whose orbit is smaller than the orbit of the earth. There are two inferior planets, Mercury and Venus. A *superior planet* is a planet whose orbit is larger than the orbit of the earth. Mars, the four major planets, and Pluto are superior planets. The apparent behavior of the inferior and superior planets is quite different. Because Mercury and Venus have orbits smaller than Earth's, they never appear very far in the

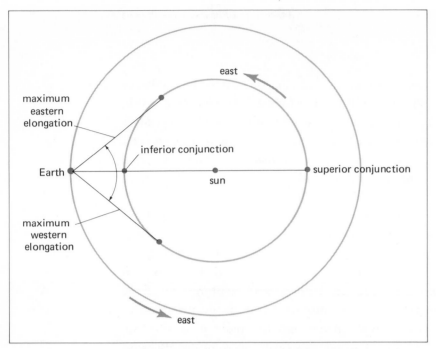

Figure 4.10 **The Aspects of an Inferior Planet**
This drawing shows the various aspects of a planet whose orbit lies inside the orbit of the earth. Only Venus and Mercury have orbits smaller than that of the earth.

sky from the sun. The Ptolemaic system accounted for this fact by having the centers of the epicycles of these two planets remain on a straight line joining the earth to the sun as those centers moved along their respective deferents.

Kepler's third law tells us that a closer planet moves more rapidly in its smaller orbit than a planet that is farther from the sun. The orbital period of Mercury is 88 days; that of Venus is 225 days. During one Earth year, Mercury completes about four revolutions, gaining three laps on the earth. Consequently, seen from the earth, Mercury appears to shuttle back and forth in an ellipse, which we see almost (but not quite) edgewise.

Consider Mercury when it is in line with the earth and sun on the opposite side of the sun from the earth. This position of Mercury, and Venus as well, is called a *superior conjunction* (Figure 4.10). Mercury has the same celestial longitude as the sun. Its *elongation* (the difference between its celestial longitude and

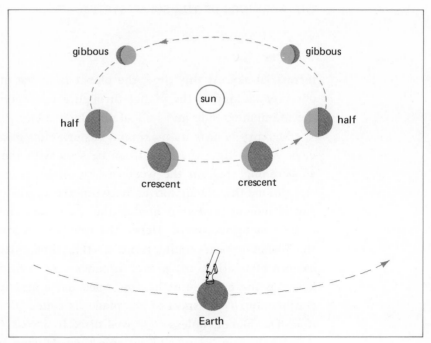

**Figure 4.11  The Phases of an Inferior Planet**
As an inferior planet moves about the sun, it exhibits phases as viewed from the
earth. The phases of Mercury and Venus can be easily seen with a small
telescope.

that of the sun) is therefore zero, but it is on the other side of
the sun from us. After passing its superior conjunction, the planet
appears to move eastward among the stars more rapidly than the
sun. Hence it is east of the sun and appears in the western sky
after sunset. Obviously, if one observes an inferior planet at this
point in its orbit with a telescope, he will see most of the illuminated half of the planet but also some of the nighttime half.
Inferior planets therefore exhibit phases, like those of the moon
(Figure 4.11). The phase that appears after a superior conjunction
is called a *gibbous phase*. It was the discovery of this fact by
early users of the telescope that disproved the part of the
Ptolemaic system which explained the motions of the inferior
planets. When an inferior planet reaches the position in its orbit
at which the line of sight from the earth is tangent to the orbit,
it has moved to what is called its *maximum eastern elongation*.
This is as far east of the sun as the planet will appear in the

terrestrial sky. At this time, the planet is in the quarter phase. An observer looking at the planet through a telescope will see half of the illuminated side and half of the dark side.

Mercury is near its maximum eastern elongation for only a few days. At this time the planet can be seen with the unaided eye, as an "evening star" in the western sky. After its maximum eastern elongation, the planet moves between the earth and the sun. The sun is moving eastward among the stars, and the planet appears to be moving westward. Hence the relative motion of the sun and the planet appears quite rapid at this time. After its maximum eastern elongation, the planet assumes a crescent shape. It comes between the earth and the sun, once more arriving at a point of conjunction. This aspect of the planet is called its *inferior conjunction*. The planet, unless it moves directly across (or *transits*) the disk of the sun, is invisible at this time. Moving from the evening sky into the morning sky, the planet again appears as a crescent, now reversed. The aspects and phases continue in reverse through the second half of the apparent revolution of the planet. The planet passes through its crescent, quarter (at its maximum *western* elongation), and gibbous phases and back to a full phase at its superior conjunction.

A planet's *synodic period* is the time it takes the planet to go from one aspect, as seen from the earth, back to that same aspect —for example, from one inferior conjunction to the next. During this period, the distance from the earth to the inferior planet changes greatly. At its superior conjunction, Venus is about 160 million miles away from the earth; at its inferior conjunction, it is only 25 or 26 million miles away. When it first appears as a gibbous planet in the evening sky after its superior conjunction, it is rather inconspicuous; about five weeks after its maximum elongation, it is the brightest object in the sky except for the sun and the moon. Its brightness varies with its distance from us, but of course is somewhat diminished by the changing phase. The moon is brightest when it is full and faintest when it is a crescent. Venus

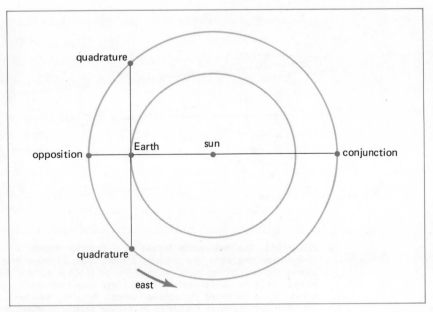

Figure 4.12 **The Aspects of a Superior Planet**
This drawing shows the various aspects of a planet whose orbit lies outside the orbit of the earth. At conjunction, the planet is invisible due to its apparent proximity to the sun. At opposition, the planet is high in the sky at midnight.

reaches its maximum brilliance as a crescent, about halfway between its maximum elongation and inferior conjunction.

The motions of the superior planets as viewed from the earth (Figure 4.12) are perhaps a bit easier to understand. Although we shall restrict the discussion to one planet for the sake of illustration, the following discussion applies equally well to all superior planets.

Let us consider Mars, and suppose that it is at *conjunction*. Mars, unlike an inferior planet, can come to conjunction only behind the sun; hence there is no need to differentiate types of conjunction. The sun moves eastward about 1° a day. Mars, due to its orbital motion, moves eastward about half a degree a day. Hence it apparently moves west of the sun, into the morning sky. Shortly after conjunction, Mars is seen in the east, before sunrise,

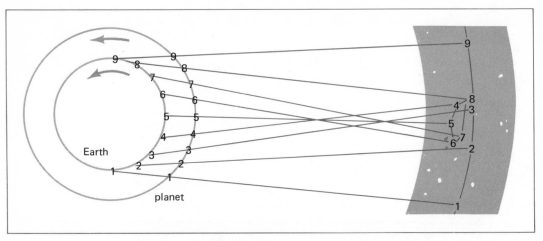

Figure 4.13 **The Retrograde Motion of a Superior Planet**
Viewed from the earth, the superior planets seem to move in a complex fashion among the background stars. Most of the time, these planets are moving eastward among the stars (direct motion). About the time they reach opposition, they appear to go backward (retrograde motion). Actually, they have been overtaken by the earth, moving more rapidly in its smaller orbit. The diagram shows the true position of a superior planet, as viewed from the earth, and its apparent position with respect to the fixed stars.

as a faint object. Its distance at that point is approximately 250 million miles. As time passes, Mars stands higher in the morning sky at sunrise (farther west of the sun). It moves eastward among the stars, but not as rapidly as the sun. After about six months Mars is in *quadrature,* 90° from the sun. A little over a year after conjunction, Mars is opposite the sun. We then say that it is in *opposition.* It rises at sunset and sets at sunrise and is highest in the sky at midnight. About the time of Mars' opposition, the earth, moving more rapidly in its smaller orbit, passes Mars, much as a race car on an inside track passes a slower competitor. Mars, as seen from the earth, appears to *retrograde,* or move westward among the stars for a while before resuming its eastward motion again, executing a loop in the sky as shown in Figure 4.13. After the retrograde loop is completed, Mars once again moves eastward, to quadrature in the evening sky. After about two years and two months, it is overtaken by the sun; it is then at conjunction again.

## 4.13 BODE'S LAW

The comparative distances of the planets have been known with some degree of accuracy since the time of Copernicus. Bode's "law," formulated in the latter half of the eighteenth century, gives a rather interesting progression of figures that closely approximate the distances of the planets. To derive Bode's law, we begin with the series of numbers 0, 3, 6, 12, 24, and so on. Except for 3, each number is obtained by doubling the preceding number. Adding 4 to each number in this series gives us another series: 4, 7, 10, 16, 28, . . . . Dividing these sums by 10 gives us yet a third series: 0.4, 0.7, 1.0, 1.6, 2.8, . . . . This final series represents fairly closely the mean distances of the major planets from the sun in AU's. The table below gives the actual distance from the sun of each of the planets and their distance as indicated by Bode's law. At the time the law was proposed, there was no known planet at a distance of 2.8 AU's, and Uranus, Neptune, and Pluto were still

**BODE'S LAW**

| Bode's Progression | Planet | Planet's Actual Distance from the Sun (in AU's) |
|---|---|---|
| (0 + 4)/10 = 0.4 | Mercury | 0.387 |
| (3 + 4)/10 = 0.7 | Venus | 0.723 |
| (6 + 4)/10 = 1.0 | Earth | 1.000 |
| (12 + 4)/10 = 1.6 | Mars | 1.524 |
| (24 + 4)/10 = 2.8 | — | — |
| (48 + 4)/10 = 5.2 | Jupiter | 5.203 |
| (96 + 4)/10 = 10.0 | Saturn | 9.539 |
| (192 + 4)/10 = 19.6 | Uranus | 19.191 |
| (384 + 4)/10 = 38.8 | Neptune | 30.071 |
| (768 + 4)/10 = 77.2 | Pluto | 39.518 |

undiscovered. Soon afterward, two things happened that seemed to strengthen the law's importance. In 1781 Uranus was discovered, and its distance (19.2 AU's) was found to coincide fairly closely with the requirements of Bode's law. In 1801 the asteroid Ceres was discovered almost exactly 2.8 AU's from the sun. The position of Neptune, which was discovered in the middle of the nineteenth century, does not coincide with the position predicted by Bode's law. (It is 30 AU's from the sun rather than 38.8.) However, Pluto, discovered in the twentieth century at 39.5 AU's, does have a mean distance close to Bode's figure of 38.8.

The significance of Bode's law has never been established. No physical reason for the relationship has ever been discovered. However, in some theories of the origin and evolution of the planets, Bode's law has been alleged to be of importance. Although there is no large planet between the orbit of Mars and Jupiter, there are thousands of minor planets, or *asteroids*, lying largely within this region. The suggestion has been made that a planet formerly occupied an orbit in this region and was shattered or, alternatively, that a planet failed to form here at the creation of the solar system, and the material which would have made up this planet instead was fractured into the asteroids. The combined mass of the total number of known asteroids is much smaller than the mass of a planet, but this does not necessarily make the suggestion invalid.

## 4.14 SIDEREAL AND SYNODIC PERIODS

As we saw in Chapter 2, the time that elapses while a planet completes one revolution around the sun with respect to a fixed point in space is called its *sidereal period*. This is the true orbital year of the planet. As viewed from the moving earth, the motion of the planets is somewhat complicated. As we noted in Section 4.12, the period from one aspect of a particular planet until it returns to that same aspect is called a *synodic period*. Although the sidereal period of Venus is about $7\frac{1}{2}$ of our months, its synodic period is about 19 months. Mars, a superior planet, has a sidereal period of 687 days and a synodic period of 780 days. The relation between the two periods, for any inferior planet, is

$$\frac{1}{\text{synodic period}} = \frac{1}{\text{sidereal period}} - \frac{1}{\text{Earth sidereal period}}$$

For a superior planet, it is

$$\frac{1}{\text{synodic period}} = \frac{1}{\text{Earth sidereal period}} - \frac{1}{\text{sidereal period}}$$

The various fractions represent the angular rate at which the planets travel with respect to the sun and, in the case of the synodic period, with respect to the earth. The first formula can be used to calculate the rate at which an inferior planet gains on the earth. The second formula can be used to obtain the rate at which the earth gains on a superior planet.

# QUESTIONS

(1) What is the purpose of epicycles in the Ptolemaic system?
(Section 4.1)

(2) Who was Nicolaus Copernicus?          (Section 4.3)

(3) Why did Tycho Brahe conclude that the earth does not go around the sun?          (Section 4.4)

(4) What do Kepler's three laws tell us, and why are they important?          (Section 4.5)

(5) What observations made by Galileo supported the Copernican system?          (Section 4.6)

(6) What contribution did Sir Isaac Newton make in the field of planetary motions? (Section 4.9)

(7) Why do the great mass of the sun and the isolation of the solar system simplify the problem of explaining planetary motions? (Section 4.10)

(8) Name the planets in order outward from the sun. (Section 4.11)

(9) What planets were *not* known to ancient man? (Section 4.11)

(10) Draw a schematic diagram of the aspects of an inferior planet. Be sure to label points of conjunction, maximum elongation, and so on. (Section 4.12)

(11) Draw a schematic diagram of the aspects of a superior planet. Be sure to label the points of conjunction, opposition, quadrature, and so on. (Section 4.12)

(12) What does Bode's law attempt to tell us? (Section 4.13)

(13) What is the difference between the sidereal period of a planet and its synodic period? (Section 4.14)

# CHAPTER 5

# THE MOON, OUR NEAREST NEIGHBOR

Of all the objects in the sky, the closest to the earth is our one natural satellite, the moon. At times during the month its brilliant light dominates the nighttime sky. It has been an object of wonder, and even veneration, in ages past. Except for the sun, it is the brightest object in the heavens. It was probably the first celestial object to be studied with the telescope, and by far the largest part of our space program to date has been devoted to exploring it.

## 5.1 THE ORBIT OF THE MOON

The moon revolves about the earth, or more properly, the earth-moon system revolves about a common center of mass, once in what is called a month. As we shall see, there are various types of months. The word *month* comes from the same root as the word *moon*, and each type of month is related in one way or another to the moon's motion. Ancient astronomers realized that the moon revolves about the earth, but it was only relatively recently that it became possible to determine the distance from the earth to the moon (that is, the dimensions of the moon's orbit) with reasonable precision. The distance of an object from the earth is usually determined by some sort of measurement of parallax. *Parallax*, as we noted in Chapter 4, is an apparent shift in the position of a nearby object with respect to a more distant background (Figure 5.1). It may be due to a shift in the position of the observer or, if there are two widely separated observers, to the distance between them. Since the moon is relatively close to the earth, parallax can be detected even when the baseline distance between two observers is fairly short. Depending on the circumstances, almost any reasonable separation will do; however, the observers' findings are usually standardized in terms of a baseline equal to the earth's radius. Observations of the moon's parallax give its distance from the earth as roughly 240,000 miles (Figure 5.2).

Each month, the distance from the earth to the moon varies from 221,463 miles to 252,710 miles, averaging 238,857 miles. The point in the moon's orbit closest to the earth is called its *perigee* (Figure 5.3). The opposite point, at which the moon's orbit is furthest from the earth, is called its *apogee*. The line joining these two points is the major axis of the elliptical orbit and is called the *line of apsides*. This line does not remain fixed in space, but turns through 360° in about nine years.

The parallax of the moon has been determined in a variety of ways. The earliest determination was made by observations at the ends of baselines of known lengths. This is the classical method; the parallax thus obtained is called the *direct*, or *trigonometric*,

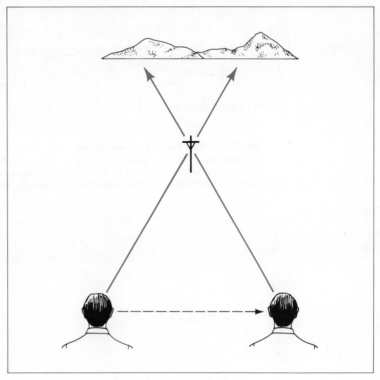

**Figure 5.1  Parallax**
To a moving observer, the position of a nearby object with respect to a distant background appears to change. This is called parallax. By measuring the straight-line distance he has moved and the apparent shift in the position of the object, the observer can calculate the distance to it.

*parallax* because it is measured directly and the distance is computed using trigonometry. The U.S. Army, in 1946, and the Office of Naval Research, in 1957, determined the distance of the moon by timing the transmission and return of radar pulses directed at it. In 1957, the Office of Naval Research announced a distance to the moon that was accurate within three-quarters of a mile. In July 1969, the astronauts of Apollo 11 deposited upon the surface of the moon several experimental packages, one of which was a laser ranging reflector (Figure 5.4). This device and the telescopes at

**103**

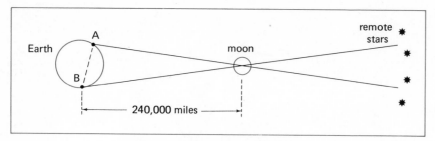

Figure 5.2 **The Parallax of the Moon**
Because the moon is close to the earth, its parallax is large. Simultaneous observations made at A and B will reveal a measurable difference in its apparent position with respect to the background of remote stars.

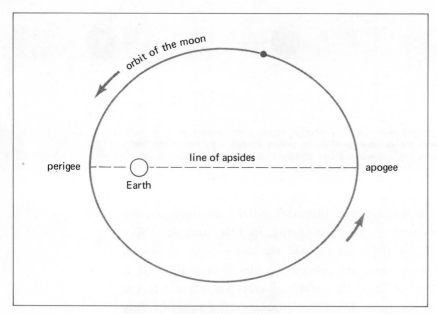

Figure 5.3 **The Orbit of the Moon**
The moon revolves, relative to the earth, in an ellipse of moderate eccentricity (exaggerated here) with the earth at one focus. It is nearest to the earth (221,463 miles away) at its perigee and farthest (252,710 miles away) at its apogee.

Figure 5.4 **The Laser Ranging Retro-reflector**
This instrument, set up on the moon by the astronauts of Apollo 11, allows precise measurements of the distance from the earth to the moon. NASA photograph.

the Lick and McDonald observatories have been used to measure the transmission time of light between the telescopes and the surface of the moon. The final uncertainty in the mean distance from the earth to the moon is expected to be about 6 inches.

The difference in the moon's distance at its apogee and perigee is a little over 31,000 miles, or 12 percent of the maximum distance from the earth to the moon. Hence the moon as seen from earth does not have a constant apparent size; its diameter at perigee is about 12 percent larger than its diameter at apogee. Since it takes the moon about two weeks to move from the closest to the farthest points in its orbit, this change is not readily noticed. However, there is another change in the apparent size of the moon which is purely psychological but which is often remarked even by casual observers. To many people, the full moon appears much larger when it is rising than it does 6 hours later when it is high in the sky. Actually, the moon as measured with a micrometer on a telescope is a little smaller when it is on the horizon than when it is high in the sky. This is because the observer is closer to the moon by one earth radius when it is at the zenith than when it is at the horizon.

## 5.2 THE RELATIONSHIP OF THE MOON'S ORBIT TO THE ECLIPTIC

The plane of the moon's orbit does not lie exactly in the plane of the ecliptic. In fact, the angle between the two planes is close to 5°. The line connecting the two points at which the moon's orbit crosses the ecliptic is called the *line of nodes* and passes through the earth (Figure 5.5). The points of intersection are called *nodes;* when the moon moves from the southern side of the ecliptic to the northern side, it is said to be at the *ascending node.* The opposite point is called the *descending node.*

The line of nodes moves through 360° once in about 18.6 years. In about 9.3 years, the two nodes change places; 9.3 years later, they return to their original position. This motion should not be confused with the motion of the line of apsides, which is a movement at the orbit in its plane rather than a movement of the plane itself. The *regression* of the moon's nodes has several results. One is a nodding effect called *nutation* (like the nutation of the earth's axis). Another is a change in the declination limit of the moon. When the ascending node is at the vernal equinox, the moon's orbit is 5° above the ecliptic at the ecliptic's northernmost point, and the maximum declination of the moon is approximately $28\frac{1}{2}°$ ($23\frac{1}{2}°$ + 5°). Nine and one-third years later, when the nodes are interchanged, the maximum declination of the moon is $18\frac{1}{2}°$ ($23\frac{1}{2}°$ − 5°).

106

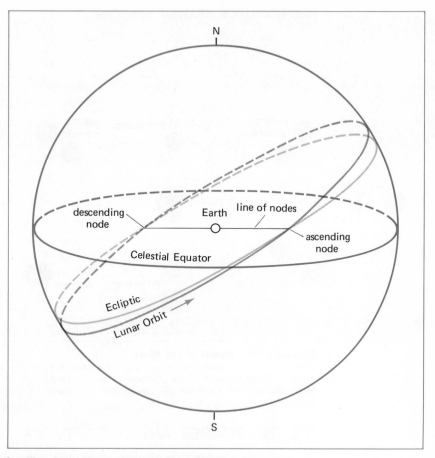

Figure 5.5 **The Relationship of the Moon's Orbit and the Ecliptic**
The orbit of the moon makes an angle of about 5° with the orbit of the earth about the sun. The line of intersection of the two planes (the orbit and the ecliptic) is called the line of nodes since it passes through the lunar nodes (where the orbit and the ecliptic apparently intersect). The line rotates once in about 18.6 years. The regression of the nodes, as this phenomenon is called, changes the apparent path of the moon slightly in successive months.

The moon's course among the constellations differs considerably from month to month as a result of the change of position of its orbit relative to the ecliptic. Nevertheless, it always remains within the confines of the zodiac.

Since the full moon is always opposite the sun and is never very far from the ecliptic in the course of a month, it is high in the sky in wintertime and low in the summer. The moon's other phases change correspondingly with the seasons of the year.

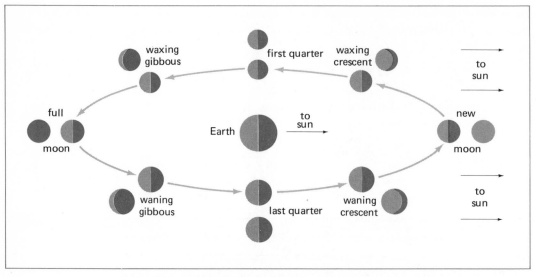

Figure 5.6 **The Phases of the Moon**
Although the moon always keeps the same face toward the earth, its
appearance changes with the position of the terminator, the boundary between
night and day on the moon as viewed from the earth.

## 5.3 THE PHASES AND ASPECTS OF THE MOON

To any observer, the most noticeable thing about the moon is the
change in its appearance from night to night (Figure 5.6). During
the lunar month the moon is first a slender crescent that appears in
the western sky, shortly after sunset. It disappears within an hour
or two after the sun goes down. From night to night the crescent
broadens and the moon sets later. About a week after the new moon,
half of the side of the moon facing the earth is illuminated and the
moon sets about midnight.

The edge of any celestial body that is the geometric edge of its
sphere is called the *limb*. In the case of the moon at first quarter,
the western curved edge is the limb. The eastern edge, which looks,
to an observer on the earth, like a straight line, is in reality a circle
viewed on edge. This circle is called the *terminator* and marks the
boundary between day and night on the moon.

108

A few nights after the appearance of the first quarter moon, the terminator begins to bulge out to the east and more than half of the visible surface of the moon is illuminated by sunlight. This is the *gibbous* phase, when more than half of the side of the moon facing the earth is illuminated. Approximately two weeks after the new moon, the earth is between the sun and the moon, and the moon is full. The full moon rises as the sun sets and is up all night, setting at sunrise. After this phase, the moon again becomes gibbous and wanes to the last quarter, crescent, and back again to new. The complete cycle takes about 29.5 days, a period that is defined as the *synodic month*.

To describe the position of the moon in the sky with reference to the sun, we use the term *elongation*. The elongation of the moon is the difference between its celestial longitude and that of the sun. In other words, it is the moon's angular distance from the sun. The celestial latitude of the sun is 0°; that of the moon may be anything from about 5°N to 5°S. An elongation of 0° means a new moon. Special elongations have special names. When the elongation of the moon is 90°, it is said to be at quadrature. This occurs when the moon is nearly a quarter (Figure 5.7). When its elongation is 180°, the moon is full. This position of the moon is called the *aspect of opposition*.

A clearer understanding of what produces the phases of the moon can be gained by a simple experiment. Suppose that a single source of illumination is placed in one corner of an otherwise darkened room and a tennis ball or any other spherical object is held at arm's length between the observer and the light. The dark side of the tennis ball is toward the observer, and the elongation is zero, corresponding to the elongation of the new moon. As the observer turns slowly to his left, he will see a slender crescent of light begin to appear along the right edge of the tennis ball. When he has turned approximately 90°, he will find that the right half of the tennis ball is illuminated and the left half is dark. As the observer continues to turn, the tennis ball will pass through a gibbous phase. When his

Figure 5.7 **The First Quarter Moon**
Eight days into the lunar month the moon is at the first quarter stage. In this picture, the south is at the top. Photograph from the Hale Observatories.

back is to the light, the entire tennis ball will be illuminated, as the moon is when it is full (Figure 5.8). Turning through another 180°, the observer will see another gibbous phase, the last quarter, a waning crescent, and finally a new "moon" again.

110

Figure 5.8  **The Full Moon**
When the moon is full, the sun is behind the terrestrial observer and the flat
lighting does not show the relief observed near the terminator at other phases.
However, the seas show well, and craters, such as Tycho (near the top) with its
ray system, are prominent. Photograph from the Hale Observatories.

During the month of the phases, a number of interesting phenomena may be observed, even with a small telescope. As the sun
rises, the mountains and other irregularities of the moon near the
terminator cast long shadows that shorten as the sun climbs higher
in the lunar sky. Since a day on the moon is the same length as a
synodic month, the change in the length of the shadows from night
to night is easy to observe. Figure 5.9 shows the major features on
the surface of the moon.

Another easily observed effect is that of earthlight on the moon.
Almost everyone has noticed this at one time or another. Frequently,

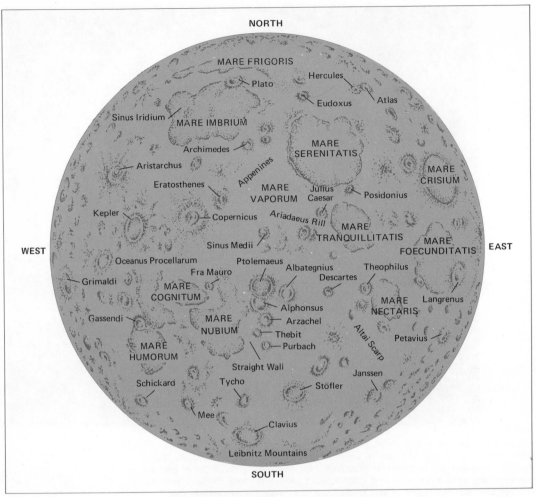

**Figure 5.9  A Map of the Moon**
Prominent features, such as craters, on the moon's surface have been named after famous scientists and philosophers. In addition, other features have received fanciful names such as *Mare Imbrium* ("the Sea of Showers") and *Mare Nectaris* ("the Sea of Nectar"). In this drawing, the north is at the top.

when the slender crescent moon can be seen in the darkening western sky, the entire moon is faintly visible. The bright crescent shows up quite well; but as the sky darkens, the rest of the moon, fainter but nonetheless visible, appears against the black of the sky. This phenomenon is sometimes referred to as "the old moon in the new moon's arms" and is caused by earthshine on the moon (Figure 5.10). The earth reflects nearly five times more sunlight than the

112

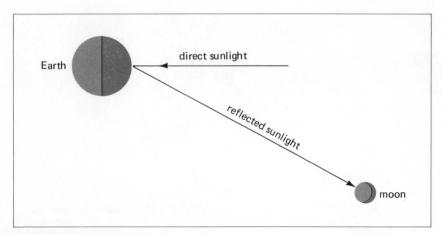

**Figure 5.10  Earthlight on the Moon**
The slender crescent before and after the new moon is not bright enough to mask the light reflected from the gibbous earth. The portion of the lunar disk hidden from the sun, otherwise in darkness, may be faintly seen.

moon, and the area of the earth is about 16 times the area of the moon; consequently, the amount of light returned from the almost full earth just after the new moon is perhaps 75 or 80 times what the earth receives from the full moon. The terrestrial landscape on a clear moonlit night is quite bright; the lunarscape under similar conditions is even more brightly illuminated.

## 5.4 THE EFFECTS OF THE MOON'S REVOLUTION

The period of the revolution of the moon around the earth with respect to a fixed direction in space (or a fixed star) is called the *sidereal month*. It is close to 27⅓ days. Thus the moon moves approximately 13° in a solar day, or a little over half a degree an hour. Its revolution is direct (counterclockwise as viewed from the north celestial pole), and it appears to move eastward among the stars by an amount equal to its own diameter in an hour. Since the moon is about 240,000 miles from the earth and completes one revolution in

**113**

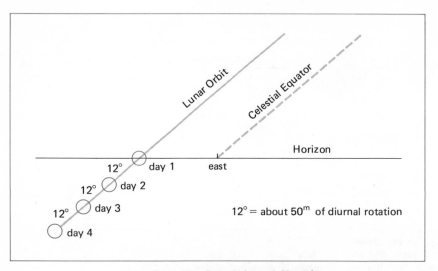

**Figure 5.11  The Retardation of Moonrise**
Because the moon appears to move eastward about 13° among the stars each 24 hours and the sun moves only about 1°, the moon is about 12° farther east of the sun at each successive moonrise. Thus the moon rises about 50 minutes later each night, on the average. The drawing assumes that the moon is in a part of its orbit approximately parallel to the celestial equator.

$27\frac{1}{3}$ days, its velocity in its orbit is nearly 2000 miles per hour. The earth itself turns eastward, causing the stars and other objects of the heavens to appear to move from east to west. However, the revolution of the moon causes it to appear to move among the stars from west to east. As a result of the combination of these two motions (of the earth and the moon), one apparent circuit of the sky by the moon takes an average of 24 hours and 50 minutes. If we kept time by the moon, what we might call a lunar day (that is, one apparent trip of the moon about the earth) would be 24 hours and 50 minutes of solar time in length. Since we keep solar time, this means that the moon rises an average of about 50 minutes later each day (Figure 5.11). However, this retardation of moonrise can depart considerably from that average. For example, around the time of an autumnal equinox, late in September, the ecliptic is inclined to the horizon at an unusually shallow angle. As a result, although the moon moves about 13° in its orbit each day, the full effect of this 13° movement is considerably diminished by the shallow inclination of the moon's orbit. The net result is that the moon appears to drop back less from night to night. Farmers noticed this phenomenon many centuries ago. The full moon nearest the autumnal equinox would be bright in the sky shortly after sunset for several days in a row. The farmers called this moon, by whose light they were able to work in the fields, the *harvest moon*. The full moon following the autumnal equinox is usually known as the *hunter's moon*.

It is easy for the casual observer to see the revolution of the moon by noting its position with respect to a particular bright star. Moreover, since the moon is closer to the earth than anything else in the sky, it frequently obscures objects behind it. When the moon appears to cover a star, this is called an *occultation*. Predictions of occultations of stars by the moon are given in various astronomical publications and may be observed with a small telescope. By accurately timing occultations, we can obtain accurate measures of the position of the moon in the sky; consequently, observations of these events are used in the study of the moon's motion.

The sidereal and synodic months differ because of the revolution of the earth-moon system about the sun. Their relationship is shown in Figure 5.12.

## 5.5 THE REVOLUTION OF THE EARTH-MOON SYSTEM

Up to now, we have considered only the moon's motion relative to the earth. However, the earth-moon system revolves about its *center of mass*. This point, for a system such as the earth and the moon, is defined as the point at which the system would balance if the two members could be connected by a rigid, weightless rod.

Both bodies move in orbits of similar shape but different sizes. Thus the center of the earth (as well as that of the moon) moves in

**115**

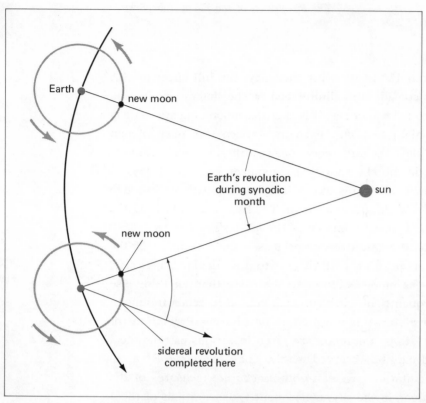

**Figure 5.12 Synodic and Sidereal Months**
The moon revolves about the earth in 27⅓ days, called the sidereal month, because as seen from the moving earth this is the period of time during which the moon appears to orbit the earth from the vernal equinox back to the equinox again. During this time the earth has moved about 27° along its orbit and the sun has, as a result, appeared to move along the ecliptic. It takes the moon about 2⅛ more days to overtake the sun and complete the synodic month of 29½ days.

an ellipse about the center of mass. The amount of this motion can be measured by observations of the other planets and asteroids. From these observations, we find that the earth is about 81 times more massive than the moon. Furthermore, the center of mass of the two bodies is approximately 2900 miles from the center of the earth, in the direction of the moon. It is not a fixed point because the earth rotates faster than the system revolves, so that the center of mass moves with respect to the geography of the earth.

More refined measurements of the moon's mass are now available from measurements of the paths of orbiting lunar vehicles and

**116**

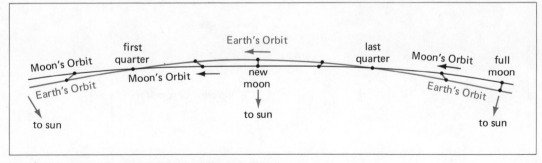

**Figure 5.13 The Paths of the Earth and Moon in Space**
Since the earth and moon revolve around their common centers of mass, they
individually follow wavy paths while the center of mass traces out what we have
heretofore called the earth's orbit. The amplitude of the oscillation in the path
of the moon is over 80 times as great as that in the earth' path. The drawing is
not to scale.

the effect of the moon's gravity upon them. We now estimate the
moon's mass at about $7.3 \times 10^{25}$ grams.

The moon follows a wavy path about the sun, but so large is
the orbit of the earth-moon system about the center of mass in
comparison with the orbit of the moon that the path of the moon is
always concave toward the sun. It is almost impossible to make an
accurate scale drawing of the true motion of the moon about the
sun because of the disparity in dimensions, although Figure 5.13
may be considered a reasonable approximation.

If we know the distance from the earth to the moon, it is possible, from an observation of its apparent angular diameter, to compute its true diameter of 2160 miles. Our moon is not the largest of
the satellites in the solar system; it is the fifth in order of size. However, in proportion to its *primary* (the object about which it revolves), the moon is by far the largest of the satellites. The earth-
moon system can almost be said to be a twin planet, and in the solar
system it is unique in this respect. If Jupiter, for example, had a
satellite that was as large compared to its primary as our moon, it
would be nearly 24,000 miles in diameter, or almost as big as Neptune. Hence we can say that we live on one member of a twin planet,

**117**

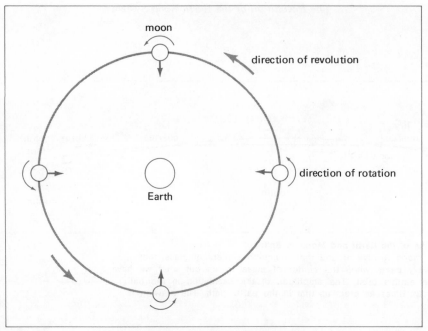

Figure 5.14 **The Rotation of the Moon**
The rotation of a celestial body is measured with reference to a fixed direction
in space (for practical purposes, the stars). Since the moon's periods
of revolution and rotation are the same, we can never see
the far side of the moon from the earth.

although, in many respects, the twins are quite dissimilar.

Knowing the mass of the moon and its size, we can compute its average density, which is 3.3 times the density of water. Surface rocks on the earth have an estimated median density of 2.6 times the density of water.

## 5.6 THE ROTATION OF THE MOON

The moon rotates upon its axis once during the sidereal month. Consequently, the same side is always toward the earth, as in Figure 5.14. Until the dream of lunar exploration became a reality, we had no idea of the appearance of what we call the far side of the moon. However, an observer on Mars, equipped with a powerful telescope, would be able to see all sides of the moon during a sidereal month.

118

**Figure 5.15  The Librations of the Moon**
In the picture on the left, Mare Crisium is closer to the limb, demonstrating
libration in longitude. The larger craters at the top (the southern region) are
closer to the limb in the picture on the right, demonstrating libration in latitude.
Lick Observatory photograph.

At times, however, we can see slightly more than the near hemisphere of the moon. This phenomenon is called a *libration* (Figure
5.15). Because the moon's rotation is nearly constant while its
revolution, due to the ellipticity of its orbit, is variable, the rotation
gets ahead of and behind the revolution. This variable relationship
between these two movements enables us to see around the
eastern and western edges of the moon at certain times of the

**119**

month. This is called a *libration in longitude*. Because the moon's axis of rotation is not perpendicular to its orbit, we also see over the north pole and the south pole. This is called *libration in latitude*. The earth's position at these times is analogous to that of the sun when it shines over the North Pole of the earth in the northern summer and over the South Pole of the earth in the southern summer, producing the phenomenon of the midnight sun. There is also a third libration called *diurnal libration*. When the moon is on the horizon, we are actually observing it from a point some 4000 miles above the center of the earth. Hence we are able to see a bit further beyond the limb than would be possible from the center of the earth. We see a bit over the limb as the moon is rising and again as it sets. These three librations make it possible for us to see a total of 59 percent of the surface of the moon at one time or another; 41 percent is never visible from the earth.

## 5.7 SURFACE CONDITIONS ON THE MOON

We have known for a long time that the moon has no atmosphere. The reason for this is quite simple. The surface gravity of the moon is considerably smaller than the surface gravity of the earth. Since the mass of the moon is $\frac{1}{81}$ the mass of the earth, one might expect its surface gravity to be $\frac{1}{81}$ as strong. However, surface gravity depends not only on the mass of the body but also on its radius. We know from the law of universal gravitation (Section 4.9) that the moon's attraction for an object on its surface varies directly with its mass and inversely with the square of its radius. As a result of its smaller mass and smaller radius, the moon has a surface gravity which is $\frac{1}{6}$ that of the earth. Thus a man who weighed 180 pounds on the earth would weigh only 30 pounds on the moon.

An *atmosphere* is an envelope of gas surrounding a celestial body. Gas consists of molecules and atoms moving about at high speeds. Their exact speed depends on the temperature of the gas.

120

The higher the temperature, the higher their velocity. Under normal conditions, these tiny particles move rapidly and collide with great frequency. Their average velocity under any particular set of conditions of temperature, pressure, and so on may be predicted, but no particle maintains that average velocity for a long time, due to the many collisions that occur. If a particle near the upper surface of an atmosphere achieves enough velocity, it can escape into outer space. This velocity is called the *velocity of escape* and is a function of the surface gravity of the body. For the earth, the velocity of escape is in the neighborhood of 7 miles per second; for the moon, it is about 1½ miles per second. Thus the earth is able to restrain most elements, except free hydrogen, from escaping, and the nitrogen and oxygen of our atmosphere are permanent residents. On the moon, the velocity of escape is so low that no atmosphere has been retained. There are various ways in which an observer could detect the existence of even a rarefied atmosphere on the moon's surface. For example, when a star or planet is occulted by the moon (Figure 5.16), particularly by the dark limb, there is no evidence whatever of a gradual dimming of the star's light, as there would be if the moon had an atmosphere. Nor is there any evidence of a lunar atmosphere during a solar eclipse.

Because the earth has an atmosphere and the moon does not, surface conditions on the two bodies differ. During the night, our atmosphere acts as a blanket, retaining much of the warmth received during the day. During the day, it insulates us from the direct rays of the sun and absorbs excess solar heat. This insulation keeps our surface temperature within a reasonable range. In fact, it is rather unusual for the temperature in the temperate zone to change during a 24-hour period by much more than 20°C.

The moon, lacking an atmosphere, experiences a greater range of temperatures during the lunar day, which lasts for a full month, and more extremes of heat or cold are possible. Observations of the lunar surface indicate that the surface temperature may exceed 100°C when the sun is at the zenith and drop to −150°C at midnight.

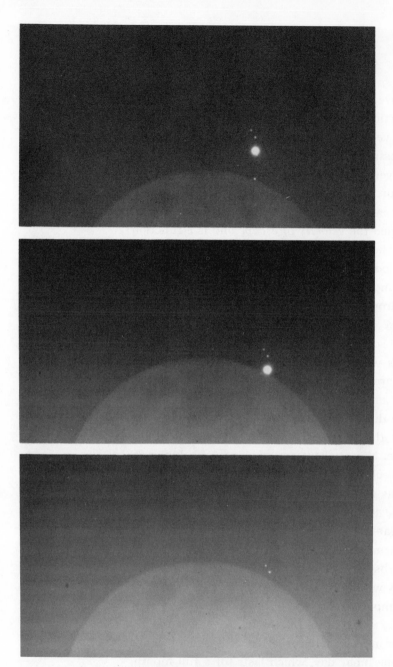

Figure 5.16 **A Lunar Occultation of Jupiter**
This series of photographs shows the emergence of Jupiter and three of its
satellites after their occultation by the moon. The sudden disappearance and
reappearance of celestial bodies during lunar occultations demonstrates that the
moon does not have an atmosphere. Griffith Observatory photographs by Paul Roques.

122

Figure 5.17 **The Crater Clavius**
This crater is the largest on the earthward side of the moon, about 150 miles in diameter. Photograph from the Hale Observatories.

During a lunar eclipse, when the moon is momentarily cut off from the direct rays of the sun, a drop of 150°C in surface temperature was recorded by devices left behind on the moon by Apollo astronauts. This indicates that the surface material of the moon has little conductivity, and the observed temperatures are probably quite superficial. In this respect as in others, the surface material of the moon resembles somewhat the rocks of the earth's crust.

The lack of an atmosphere on the surface of the moon is partly responsible for its low reflectivity, or *albedo*. The albedo of an object is simply the percentage of light it reflects. The moon reflects approximately 7 percent of the visible light it receives from the sun. Consequently, we say that the moon's albedo is 0.07.

The lack of a protective blanket of atmosphere also means that meteorites strike the surface of the moon by the millions every day, sometimes producing large craters (Figure 5.17). In the days before the first lunar landings, it was thought that portions of the moon might be covered with a thick layer of meteoric dust that might in

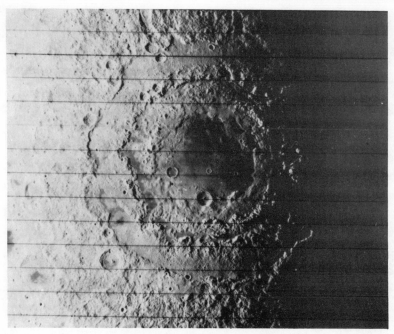

Figure 5.18 **The Basin Orientale**
This formation is just out of sight around the eastern limb of the moon as viewed from the earth. In this picture, the south is at the top. The Cordillere Mountains form the outer scarp, a little over 600 miles in diameter. The texture and sharpness of the mountains suggest that this is probably one of the youngest of the large lunar basins. NASA photograph.

some cases be miles deep. Fear was expressed that the dust could swallow a lunar landing vehicle without a trace. Up to now, however, there has been no evidence of such deposits of dust on the moon's surface.

## 5.8 THE LUNAR SEAS

The lunar seas are the dark areas on the moon's surface readily visible to the unaided eye. They are called seas, or *maria*, because ancient astronomers thought they were bodies of water or the beds of dried-up seas. Some bear fanciful names, such as Mare Serenitatus ("the Sea of Serenity") and Sinus Iridum ("the Bay of Rainbows"). The seas are concentrated mostly in the northern half of the hemisphere facing the earth and seem, with one exception, to be inter-

124

connected. They are generally circular and may be the result of the impact of extremely large meteorites in the early history of our moon. They are dark because they are smooth and because the particles covering them reflect less than the average amount of light. There is only one sea on the far side of the moon (Figure 5.18).

The notion that the seas may have been formed by the impact of meteorites has been strengthened by the discovery of irregular concentrations of mass in the regions of some of the seas, which cause slight perturbations in the orbits of satellites circling the moon. This information is one of the many direct results of our space program. However, it is still impossible to say definitely exactly how the seas and other features on the surface of the moon were formed. The seas appear to be smoother than the rest of the moon and may represent large plains of dark lava. Certainly, some of the formations on the moon are igneous, and the presence of lava on the surface should not be surprising.

## 5.9 CRATERS

A good share of the near side of the moon and most of the far side is covered with millions of *craters*. Those that bear names have been named after eminent scientists, philosophers, and historical figures such at Plato, Archimedes, and Copernicus (Figure 5.19). On the near side of the moon there are about 30,000 craters large enough to be seen with terrestrial telescopes. Orbiting satellites and photographic observations of the entire surface of the moon have revealed that there are actually millions of craters. They range in size from over 150 miles in diameter down to tiny craters so small that a microscope is needed to inspect them.

There are two basic theories about the formation of craters, the volcanic theory and the meteorite theory. What looks like volcanic activity has been observed on the lunar surface. However, most of

Figure 5.19 **A Closeup of Copernicus**
This picture was taken by Lunar Orbiter II from a point 28 miles above the moon's surface and about 150 miles due south of the center of the crater Copernicus. The mountains rise as much as 1000 feet, with slopes of up to 30°. NASA photograph.

the craters resemble impact craters rather than volcanic ones, since the volume of displaced material is about equal to the estimated volume of the surrounding walls. Then too, the craters are for the most part broad and shallow, as impact craters would be. On the other hand, the floors of some have been filled or partially covered with what appear to be lava flows, and in many instances there is a central peak or peaks.

The lack of similar craters on the earth, which might at first glance seem to support the volcanic theory, actually does not, since the surface conditions on the earth and moon are not comparable.

The earth has an atmosphere and weather. The moon does not. Fewer meteorites can reach the surface of the earth, and those that do have been slowed down by the atmosphere. Moreover, craters formed when the earth was young would for the most part have been worn away long ago by erosion, weather, scouring by glaciers, and so on. Nevertheless, there are craters on the surface of the earth obviously made by meteorites, and many of them resemble the craters of the moon. Moreover, photographs of Mars taken by space vehicles in the Mariner series show unquestionably that certain areas of the planet closely resemble the cratered surface of the moon. Consequently, there is some evidence that cratering activity may have occurred at one time throughout our particular region of the solar system.

## 5.10 MOUNTAINS, RAYS, AND RILLS

The mountain ranges on the moon are lofty indeed considering the moon's small diameter. Some of them, at the edges of the great seas, rise as high as 18,000 feet, and the Leibnitz Mountains near the south pole are 26,000 feet from base to peak. Mountains on the earth are measured from sea level; since no comparable reference point exists on the moon, a direct comparison of terrestrial and lunar ranges is difficult. Moreover, a mountain-raising force on the moon's surface must overcome only one-sixth the resistance it would encounter on the surface of the earth.

The lunar mountain ranges on the earthward side, such as the Alps (Figures 5.20 and 5.21), are generally named after mountain ranges on the earth. The lack of an atmosphere on the moon and the starkness of the landscape give an exaggerated impression of the ruggedness of lunar mountains. Photographs taken from lunar orbiting satellites, as well as the views obtained by the astronauts, indicate that some of our ideas about the surface of the moon, particularly the ruggedness of the landscape, are erroneous.

Figure 5.20 **The Alps**
The Alps rise above the lunar plane to the northwest (lower left) of Mare Imbrium.
The crater Plato is near the bottom; another crater, Archimedes, is at the top.
Lick Observatory photograph.

Around some of the craters, a ray pattern is easily visible. The craters Tycho and, to a lesser extent, Copernicus have light-colored streaks radiating from them. Because these show up most strongly when the sun is most nearly overhead and cast no appreciable shadows when the sun is low in the lunar sky, they are believed to be streaks of light-colored ejecta from the impact that formed the crater rather than ridges or other geographic formations. It is Tycho's ray system that makes the full moon resemble an orange.

128

Figure 5.21 **The Alpine Valley**
This photograph from Orbiter IV shows the lunar Alpine Valley, which is over 75 miles long and is aligned radially with the center of Mare Imbrium. At the time this picture was taken, the spacecraft was 1800 miles above the moon. NASA photograph.

The moon's lack of an atmosphere and its low gravity make it possible for material ejected from a crater to travel hundreds or even thousands of miles before falling to the surface. For this reason, some of the ray systems are surprisingly long.

Of quite a different nature are the rills in the surface of the moon. These long trenches, sometimes straight and sometimes winding, are almost randomly distributed over the lunar surface. What caused them is unknown, although there are several theories about their origin.

## 5.11 A SHORT HISTORY OF LUNAR EXPLORATION

One of the earliest series of lunar vehicles was the Ranger series (started in 1964). The program was plagued by initial problems, but its later record was spectacular. The Rangers, designed to crash-

Figure 5.22 **Surveyor III and Apollo 12**
Astronaut Charles Conrad inspects the Surveyor III spacecraft on the Ocean of
Storms, November 1969. Surveyor III landed on the moon on April 19, 1967. On
the horizon is the "Intrepid," the vehicle which carried Conrad and fellow-astronaut
Alan Bean to the moon's surface. NASA photograph.

land on the moon, sent out almost instantaneous photographs up
to the moment of impact. At one stage successive photographs were
used to make a motion picture film. It was jerky and perhaps not
as sophisticated as one might wish, but it was a pioneer breakthrough. The shots taken just before the crash showed craters that
were no more than a yard in diameter.

The Ranger series was followed by the Surveyor program (1966).
This vehicle (Figure 5.22) did not crash-land on the moon; it
made a soft landing, sent back pictures, dug into the moon's surface
with a small scoop, and radioed information on the structure and
characteristics of the lunar soil to scientists on earth. At this time,
there were still considerable misgivings about the safety of landing

on the lunar surface because of the possibility of areas of soft powdery dust in which a spaceship would sink out of sight. However, Surveyor landed, its tripod withstood the shock, and the vehicle did not sink out of sight. Its cameras successfully sent their message back to the earth.

Another series of lunar spacecraft, the Orbiter series, neither crashed nor made a soft landing. Instead, the vehicles orbited and photographed the moon. Orbiter 1 went to the moon in August 1966. Among the pictures it sent back was one of the surface of the moon with the earth rising, one-quarter illuminated, in the lunar sky. Orbiter 2 took the now-famous picture of the crater Copernicus from a distance of 28 miles above the surface. Orbiter 4 photographed the enormous Oriental Basin. In all, this was a very successful series of lunar vehicles.

Next came the magnificently successful Apollo series. After a number of experiments in earth orbit, the Apollo program of lunar exploration got under way. Apollo 8 circled the moon without landing. Apollo 9 and Apollo 10 were experimental. Finally, on July 20, 1969, the Apollo 11 lunar module landed on the moon, carrying astronauts Neil Armstrong and Edwin Aldrin. (Figure 5.23 shows Col. Aldrin after the planting of the U.S. flag on the lunar surface.) The two men deployed various experimental packages on the moon and collected samples of rocks from the surface.

Apollo 11 was followed by Apollo 12, which also brought back rock samples, among them a small stone containing 20 times more radioactivity than any other lunar sample and believed to be about 4.6 billion years old, or approximately the age of the earth. Apollo 13 experienced mechanical difficulties on the way to the moon and was unable to land, but the remaining Apollo missions were enormously successful. They provided geologists in laboratories around the world with hundreds of pounds of rocks, which will be studied in the minutest detail for many years to come. From such studies, we hope to gain a clearer picture of the origin and evolution of the moon, and perhaps even the solar system.

Figure 5.23 **The Flag on the Moon**
This picture of astronaut Edwin Aldrin and the flag was taken by Neil Armstrong, the first man to step on the moon, on July 20, 1969. The lunar module is on the left, and footprints can be seen in the foreground. NASA photograph.

## 5.12 THE TIDES

The vertical rise and fall of the waters of the earth twice in a little more than a day has long been associated with the motion of the moon. The interval between successive high tides at any point along the coast averages about 12 hours and 25 minutes. The complete tidal cycle takes about 24 hours and 50 minutes of solar time, or approximately as long as one complete diurnal movement of the moon around the earth. Actually, however, tides are created in the oceans by both the sun and the moon.

According to Newton's law of gravitation (Section 4.9), the strength of the gravitational attraction of one object for another varies inversely with the square of the distance between them. Thus

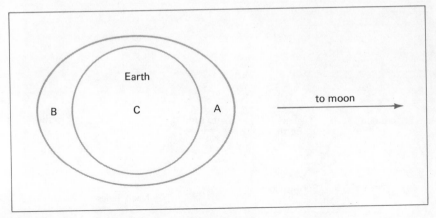

**Figure 5.24 What Causes Tides**
If the earth were covered by a deep, frictionless ocean, the moon would cause a
bulge in the water on the side of the planet nearest it and a similar bulge on the
opposite side. Although the bulge is greatly exaggerated here, for the sake of
illustration, and the actual results of the moon's attraction for the earth are
considerably more complicated, lunar tides are caused, essentially, by the difference
in the strength of the lunar gravitational field at points A, B, and C.

that part of the earth nearest the moon is attracted more strongly
by its gravitational pull than the side of the earth that is farther
from the moon. It is this *differential attraction* that produces tides.
As Figure 5.24 shows, the attraction of the moon for the water at
point A is greater than its attraction for a similar mass at the center of the earth (point C). In the same way, the moon attracts the
water at point B less than it attracts the central core. The impression of the force of attraction at A accelerates the water in the direction of the moon. The force at C accelerates the earth as a whole,
but to a slightly lesser degree. Since point B is farther away than
point C, it accelerates still less in the direction of the moon. The
result is two bulges in the water of the earth, one directly underneath the moon and one on the opposite side of the planet. These
bulges, where the level of the ocean is higher than average, are
called *high tides*. As the earth rotates upon its axis, these tidal
bulges attempt to follow the moon, or to stay beneath it. However,
due to friction in the water and on the ocean floor, there is a lag be-

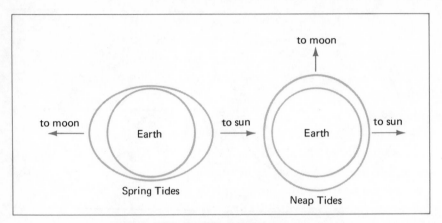

**Figure 5.25 Spring and Neap Tides**
At the time of the full and new moons, the lunar and solar gravitational fields reinforce each other, causing the most extreme tides. At the time of the quarter moon, the lunar high tide and solar low tide coincide and vice versa. Thus their combined effect is least pronounced at this time.

tween the time the moon crosses the meridian at a particular port and the high tide immediately following.

Since the sun's gravitational field is much stronger than the moon's, one might think that solar tides would be much higher than lunar tides. However, this is not the case. It is the differential force across the earth's diameter that produces tides. The sun is so far away (compared to the moon) that the difference in the acceleration of the two sides of the earth due to the sun's gravity is much smaller than in the case of the moon. There are solar tides, but the solar tide-raising force is only 40 percent of the lunar tide-raising force. At the time of the new and full moons, the lunar and solar tides are superimposed upon each other (Figure 5.25). This causes what we call *spring tides*, which are usually much higher at their highest point than average and much lower at their lowest point. At the time of the quarter moon, the two tides are at right angles to each other; the low tides are then superimposed upon the solar high tides and vice versa. Consequently, the tides at this time of the month, called *neap tides*, are less extreme.

Tide tables are prepared for the ports and coasts of the world by a number of governments. From these tables, which generally give detailed information for key locations all over the earth, the tide at almost any place upon the world's coastline can be computed. The tables are published annually, in advance, and are carried by all oceangoing vessels.

In addition to the tide in the ocean, there is also an *earth tide*. The earth tide is a deformation of the earth itself due to the same force that produces ocean tides. Since the earth is more rigid than the waters of the ocean, earth tides amount to only a few inches. However, the earth's atmosphere, being fluid, is tidally distorted in much the same way as the water of the ocean.

The horizontal motion of water over the surface of the earth causes friction. This friction results in a steady slowing of the rotation of the earth. In any isolated system, such as that of the earth and moon, energy can be transferred from one member to another, although it cannot be created or destroyed. Tidal friction causes a transfer of the energy of terrestrial rotation to the moon. This transfer causes the revolution of the moon to speed up. However, the moon is slowly spiraling outward, away from the earth. As its distance from its primary increases, its period of revolution also increases. Thus the solar day and the month are both lengthening. Eventually, the day will lengthen faster than the month, and the two will become equal to 47 of our present solar days. At some future date when the moon has spiraled outward to a certain point, the system will stabilize and the earth will turn one face toward the moon. Solar tides operating on the system will then reverse the process, causing the moon to spiral inward toward the earth.

## 5.13 ECLIPSES OF THE SUN AND MOON

Being solid spherical bodies, both the earth and the moon cast shadows in the sunlight. Because the sun is much larger than either

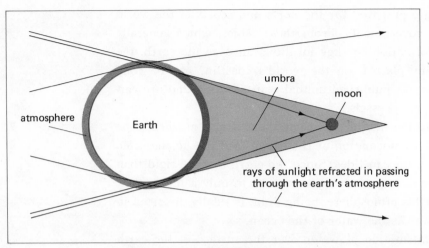

**Figure 5.26  The Moon in the Earth's Shadow**
Since the earth's shadow at the distance of the moon is over twice the moon's
diameter, a central or nearly central passage through that shadow might be
expected to darken the moon completely. However, the atmosphere around the
earth at the base of the shadow refracts light into it and selectively absorbs blue
and violet light. Hence the totally eclipsed moon is usually reddened. The arrows
indicate the refracted (and reddened) light.

the earth or the moon, these shadows take the form of tapering
cones. The shadow of the earth averages 860,000 miles long; the
moon's shadow is about 232,000 miles long. At the average distance
of the moon from the earth, the tapering terrestrial shadow is approximately 5700 miles in diameter.

In revolving around the earth, the moon passes through the
earth's shadow and a lunar eclipse occurs (Figure 5.26). Clearly,
lunar eclipses can occur only when the moon is full. Furthermore,
they must occur near the nodes of the moon's orbit; hence we do
not have an eclipse of the moon every time it is full, but at two approximately opposite times of the year, when the full moon is near
one of the nodes. If the eclipse occurs exactly when the moon is on
a node, the moon will move directly through the center of the
earth's shadow, and the period of totality (that is, the time of the
moon's complete immersion in the cone of the earth's shadow) may

136

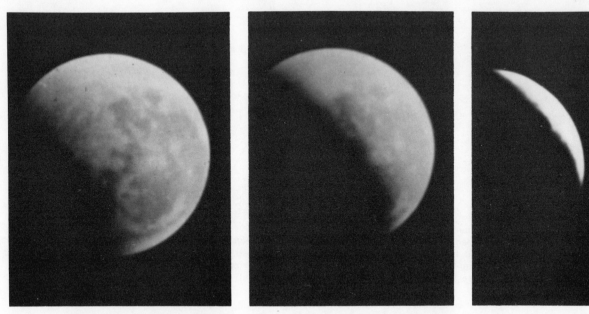

Figure 5.27 **The Moon in Eclipse**
This photograph shows three partial phases of the lunar eclipse of May 2, 1920.
The circular shadow indicates that the earth is a sphere. Yerkes Observatory
photograph, University of Chicago.

be as long as 1 hour and 40 minutes. If the moon does not pass
through the center of the earth's shadow, the period of total eclipse
will be correspondingly shorter. If the moon is far enough away
from the shadow's center, only a partial eclipse will occur, as in
Figure 5.27.

What we have called a shadow is technically referred to as an
*umbra*. Outside it, in an expanding cone, is a *penumbra*, or area of
partial shadow. When the moon passes through the earth's penumbra, there is a slight, but not very noticeable, darkening of the surface. Penumbral eclipses would not be noticed by a casual observer
unless he was forewarned of their occurrence.

Lunar eclipses are visible, weather permitting, from the entire
nighttime half of the surface of the earth facing the moon. Although
there is no particularly important astronomical information to be

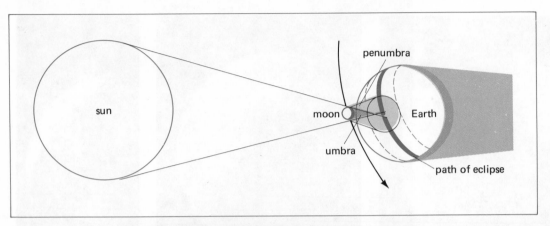

Figure 5.28 **An Annular Eclipse of the Sun**
When the apparent angular diameter of the moon is less than that of the sun and the moon passes directly in front of the sun, the result is an annular (ring-formed) eclipse. Before and after either a total or an annular eclipse (and also at places off the central track of the eclipse), the eclipse is seen as partial.

obtained from the observation of a lunar eclipse, it is an interesting phenomenon to see. As the moon first moves into the earth's umbra, a small circular area on the eastern limb of the moon darkens. This darkened area grows gradually larger, until, in about an hour, the entire moon is immersed in shadow. It remains totally eclipsed for a maximum of 1 hour and 40 minutes. The moon is not completely dark during this time, however. The atmosphere of the earth around the base of the shadow refracts light into the umbra so that, particularly at the distance of the moon and beyond, refracted light produces some illumination. Because the atmosphere is selective in its absorption of sunlight, more red light than blue is present in the umbra. Thus the moon during a typical total lunar eclipse has a coppery or reddish hue.

Solar eclipses may be total, annular, or partial. They occur at the time of new moon and, like lunar eclipses, when the new moon is at or near the nodes. As a result, solar eclipses often occur about two weeks before or after a lunar eclipse. In an *annular eclipse* (Figure 5.28), a ring, or annulus, of the sun surrounds the dark

Figure 5.29 **A Total Solar Eclipse**
The photograph shows the total solar eclipse of March 7, 1970. The inner corona is typical of the shape of the eclipse near its maximum. Photo by R. Shirkey, State University of New York at Albany.

disk of the moon. Because the average length of the lunar shadow is shorter than the average distance of the moon from the earth, *annular eclipses*, which occur when the apparent size of the moon is less than that of the sun, are more common than total eclipses. A total eclipse (Figure 5.29) occurs only when the moon's shadow is longer than average or the moon is closer to the earth than average. Under the most favorable conditions, the tip of the shadow may be as much as 167 miles wide where it touches the earth's surface (Figure 5.30); under unfavorable conditions, it may be too short to reach the surface of the earth at any point.

During a total solar eclipse, like that which occurred on June 30, 1973 (Figure 5.31), the tip of the lunar shadow sweeps across the surface of the earth. Since the shadow moves about 2100 miles per

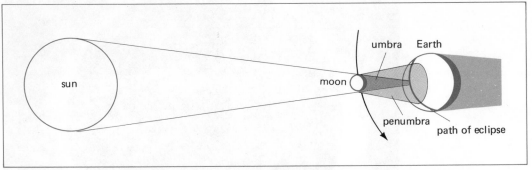

Figure 5.30 **The Lunar Shadow on the Earth**
When conditions are right, the shadow of the new moon reaches the earth, causing a solar eclipse. The path of the shadow on the earth's surface is a product of the revolution of the earth and the moon and the diurnal rotation of the earth. The moon's shadow moves at upward of 1000 miles per hour over the earth's surface and may be as much as 167 miles wide.

hour eastward over the earth's surface and the earth rotates at approximately 1000 miles an hour along the equator, the speed of the shadow at the equator when the sun is overhead is about 1000 miles an hour. The speed of the shadow may be as great as 5000 miles an hour when the sun is near the horizon. Under the most favorable conditions, the maximum duration of the total phase of the eclipse is 7½ minutes.

Eclipses may be predicted with any desired degree of accuracy. Oppolzer's *Canon der Finsternisse*, for example, predicts and charts the geographic locations of solar eclipses from 1208 B.C. to A.D. 2163. Eclipses are also listed in various almanacs, and the U.S. Naval Observatory publishes advance circulars showing definitively the paths of coming eclipses.

A total solar eclipse is one of the grandest spectacles of nature. About an hour before totality, the moon first touches the limb of the sun. During the next hour, the moon moves eastward by an amount equal to its own diameter, covering more and more of the sun, until finally only a slender crescent remains. During the latter part of the hour, the quality of the light coming from the sun changes, becom-

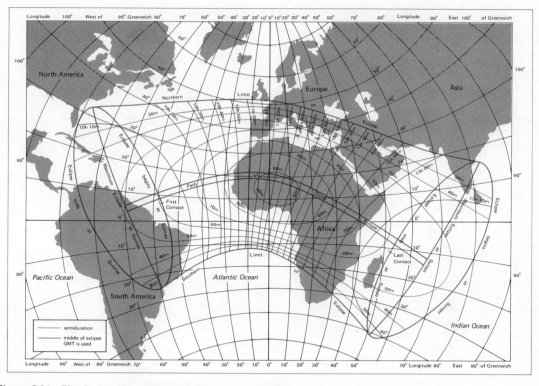

Figure 5.31 **The Path of the Solar Eclipse of June 30, 1973**
The path of totality is shown and the area of partial eclipse is outlined. Near the totality path the degree of partial eclipse is high; as the distance from the path increases, less of the sun is covered.

ing redder and attaining an eerie quality, while an unusual pallor seems to pervade the sky. Animals become disturbed, and in the last seconds before totality, *"shadow bands"* seem to run and squirm across the landscape. As the partial phase of the eclipse ends, the chromosphere, part of the solar atmosphere, is visible for a few seconds, but disappears when the moon completely covers the sun. At that time, only the sun's outer atmosphere, called the *corona*, is visible. As the total eclipse ends, the same phenomena are seen in reverse.

Astronomers can learn many things from observing solar eclipses. For example, Einstein's general theory of relativity tells us that light passing a massive body does not follow a straight line, but is deflected into a curved path. We can detect this distortion in

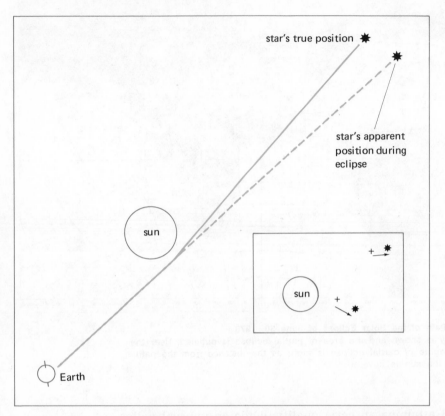

Figure 5.32 **The Relativistic Bending of a Light Ray in a Gravitational Field**
During an eclipse, a beam of starlight grazing the sun is bent. Thus stars near the sun appear to be displaced outward from their true position.

the path of starlight passing near the sun by the apparent displacement in the position of nearby stars during an eclipse (Figure 5.32). This effect constitutes one of the experimental proofs of Einstein's theory.

The number of eclipses during any one year varies from two (in which case both are solar) to seven (two lunar and five solar or three lunar and four solar). A cycle of similar eclipses, called a *saros*, recurs at intervals of about 18 years (which is about the time it takes for the line of nodes to move through 360°).

# QUESTIONS

(1) Draw a schematic diagram of the orbit of the moon about the earth. Indicate the perigee, apogee, and line of apsides.

(Section 5.1)

(2) What is meant by the regression of the moon's nodes?

(Section 5.2)

(3) Why does the moon go through phases?     (Section 5.3)

(4) What is the harvest moon? At what time of year does it occur?     (Section 5.4)

(5) Why might the earth-moon system be called a "twin planet"?     (Section 5.5)

(6) Describe the rotation of the moon. Why do we not see all sides of the moon?     (Section 5.6)

(7) What makes it possible for us to see almost 60 percent of the moon's surface from the earth?     (Section 5.6)

(8) How did astronomers in the days before lunar exploration begin to know that there was no atmosphere on the moon?

(Section 5.7)

(9) Describe the lunar seas, or maria.     (Section 5.8)

(10) Why are there many more craters on the moon than on the earth?     (Section 5.9)

(11) When did man first set foot on the moon? Describe some of the programs that led to this achievement.     (Section 5.11)

(12) How are neap and spring tides different?

(Section 5.12)

(13) Why can we still see the moon in the middle of a total lunar eclipse? What color is the moon at that time and why?

(Section 5.13)

(14) What is the difference between a total and an annular solar eclipse?     (Section 5.13)

# CHAPTER 6
# PLANETS AND SATELLITES

Up to this point, we have limited our discussion of the solar system to a description of its general characteristics. Only the moon has been examined in detail. We now turn to a study of the other planets and satellites that are members of the sun's family.

Astronomers have learned much about the planets other than the earth from using large terrestrial telescopes. Earth-orbiting satellites bearing optical equipment have provided even more information. The spectacular Mariner probes have given us a wealth of data about Mars, and Jupiter has been discovered to be the source of perhaps twice as much electromagnetic energy as can be accounted for by the reflection of solar energy alone. Tables summarizing the available data on the members of the solar system are given in Appendix 1.

## 6.1 MERCURY

Mercury is the closest planet to the sun, at a mean distance of approximately 0.4 AU. Its position can be determined by consulting an almanac or handbook. It is most visible on spring evenings, when it is near its maximum eastern elongation, or on autumn mornings, when it is at its maximum western elongation. When Mercury is east of the sun, it remains above the horizon for a short time after sunset. In the spring, the ecliptic (and consequently Mercury's orbit) is most nearly perpendicular to the western horizon at sunset, and the 28° elongation places Mercury high in the sky (Figure 6.1). In the autumn, however, the ecliptic and Mercury's orbit are low, nearly parallel to the evening horizon, and hence elongation does not place Mercury as high in the western sky. On autumn mornings, however, the situation is reversed; the ecliptic is high, and it is easy to see Mercury in the east if one knows when and where to look.

Mercury is a small planet, about 3000 miles in diameter, and its physical characteristics have frequently been compared to those of our moon. Its radius is small, and its mass is less than 5 percent that of the earth; hence its velocity of escape is about 2.6 miles a second. Its proximity to the sun gives it a high surface temperature. This, combined with a low velocity of escape, leads us to expect little or no atmosphere there, and none has ever been detected. Mercury's surface gravity is about 0.4 that of the earth.

For years, it was thought that Mercury rotated and revolved about the sun in the same period, much as the moon does about the earth. Recent radar investigations have shown, however, that Mercury's rotation period is just under 59 days. Interestingly enough, this is two-thirds of its sidereal period. Its rotation is direct, like Earth's, from west to east.

Except for Pluto, Mercury has the most eccentric orbit of the sun's nine planets. Its sidereal period is 88 terrestrial days, and its synodic period is 116 days. Seen from the earth, Mercury appears to repeat its configurations about every four months, or approximately three times per year. As we would expect from Kepler's

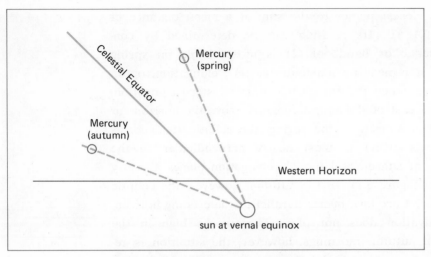

**Figure 6.1** **Mercury as an Evening Star**
The orbit of Mercury is fairly close to the ecliptic, which in springtime at sunset is above (north of) the celestial equator in the western sky. Consequently, Mercury's maximum elongation of about 28° places the planet higher in the west and makes it more easily observable. In the autumn, however, the ecliptic is below (south of) the celestial equator. Hence, even at maximum elongation, Mercury is so near the horizon as to be difficult to see at sunset. The entire situation is reversed on autumn mornings, when Mercury can easily be observed before sunrise.

third law, Mercury is also the fastest-moving of the planets, speeding around the sun nearly twice as fast as the earth.

When Mercury reaches an inferior conjunction near one of its nodes, it transits the disk of the sun (Figure 6.2). Transits of Mercury are relatively frequent and always occur in May or November. The most recent transit was in 1973, and another is scheduled for 1986. When it transits the sun, Mercury appears as a tiny black dot. Its progress is best observed by projecting the image of the sun on a white screen with a small telescope and observing the moving black dot of the planet.

Because Mercury has an eccentric orbit, the position of its *perihelion* (the point at which it is closest to the sun) is easily determined. All orbital elements in the solar system change

146

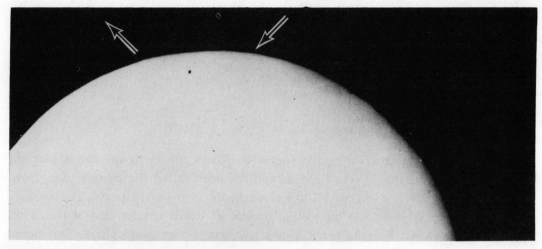

Figure 6.2 **The Transit of Mercury Across the Sun**
The arrows indicate the position of the planet with respect to the disk of the
sun and its apparent path. This picture was taken on November 14, 1907.
Yerkes Observatory photograph, University of Chicago.

slowly because of the gravitational effects of the planets on one
another. The changes are predictable and can be explained by
Newtonian mechanics. However, there is one exception to this
generalization in connection with Mercury's orbit. According to
Newtonian mechanics, the planet's perihelion point should *advance*
(move eastward) in the orbital plane 531 seconds per century.
Observation shows that Mercury's perihelion does advance, but
by 574 seconds per century. At one time it was hypothesized that
an unknown planet (Vulcan) revolving inside Mercury's orbit was
responsible for the observed excess of 43 seconds per century.
However, Einstein's theory of relativity, which, as we noted in
Chapter 5, predicts the bending of light rays by the sun during a
solar eclipse, also predicts an advance of Mercury's perihelion by
exactly the amount observed. This is regarded as evidence of the
validity of the theory of relativity. It has also been suggested that
some of this advance could be explained with Newtonian equations
if the sun were slightly oblate. However, as no really satisfactory

measure of the oblateness of the sun exists, the advance of the perihelion of Mercury is generally explained in terms of the theory of relativity.

## 6.2 VENUS

Because it is nearly the same size as the earth and has 82 percent its mass, Venus has sometimes been called the earth's twin. However, the "twins" differ markedly in many respects. The velocity of escape at the visible surface of Venus is near that of the earth, and like the earth Venus has retained an atmosphere. This atmosphere, however, is heavy and opaque; according to the 1967 Soviet probe Venera 4, it is about 90 percent carbon dioxide. Venera 4 also found that the pressure near the surface of Venus may be as much as 20 times the atmospheric pressure on the surface of the earth. Later probes gave figures as high as 80 times the atmospheric pressure of the earth. The atmosphere of Venus is also heavily clouded and gives the planet an albedo of about 75 percent. Changing markings appear in ultraviolet photographs of the planet (Figure 6.3), probably in the clouds of the atmosphere. No one has ever seen the surface of Venus, although it has been mapped by radar and radio waves (Figure 6.4).

A lack of surface reference points made the rotation period of Venus a mystery for many years. However, recent radar investigations indicate that it rotates in a retrograde direction in 243 terrestrial days. This makes it the only planet in the solar system known to have a clearly retrograde rotation, although Uranus, since its axis lies nearly in the plane of its orbit, can be considered to be rotating in either direction, depending on which pole is chosen as the porth pole. If the present figures are correct, the day on the planet Venus (measured with respect to the sun) is equal to 116 terrestrial days.

Figure 6.3 **Venus in Blue Light**
Venus' surface features are not visible; any markings that are seen are transitory
and are probably caused by clouds in the heavy atmosphere. Photograph from the
Hale Observatories.

Venus has a sidereal period of 225 terrestrial days and a synodic
period of approximately 19 months. The synodic period is the
time it takes Venus to gain one lap on the earth, and its length
is immediate evidence that the periods of revolution of the two
planets are fairly close. The orbit of Venus is the most circular
in the solar system and is inclined to the ecliptic by less than 4°.
Like the moon, Venus has phases. These are described, briefly, in
Figure 6.5.

Transits of the sun by Venus occur at inferior conjunctions

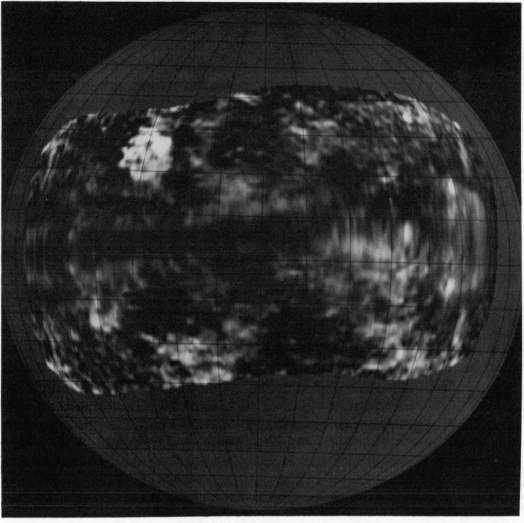

Figure 6.4 **A Radar Map of Venus**
The largest radar map of Venus was plotted by R. Goldstein and H. Rumsey, Jr., from the Goldstone, California, Tracking Station (in September 1970). The bright spot in the southern (upper) hemisphere has been tentatively identified as a mountain range. NASA photograph.

as do those of Mercury, but they are rare. The next transit will occur in the year 2004.

The heavy atmosphere of Venus is responsible for the high temperature on the surface, which is probably as much as 400°C. Heavy, opaque atmospheres produce what is called a *greenhouse*

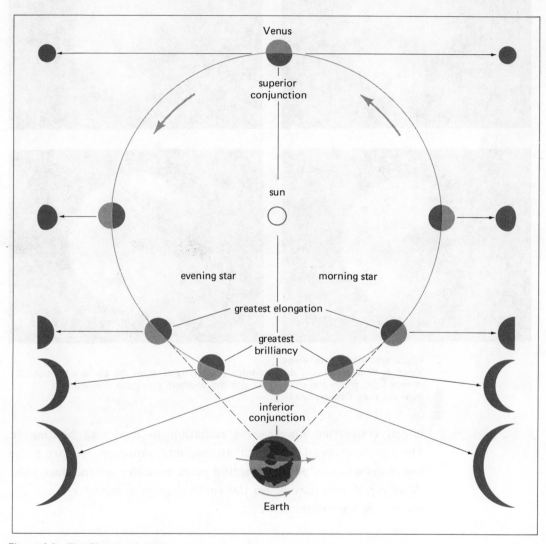

Figure 6.5  **The Phases of Venus**
Venus is about 160 million miles from the earth at its superior conjunction and
only 25 million at its inferior conjunction. As a result, its apparent size changes by a
factor of about six between the two conjunctions. Thus Venus, unlike the earth's
moon, is brightest as a crescent rather than when it is full. Venus is most brilliant
when it is about halfway between its greatest elongation and inferior conjunction.
For five weeks before and after each inferior conjunction, it is the brightest object
in the sky except for the sun and moon.

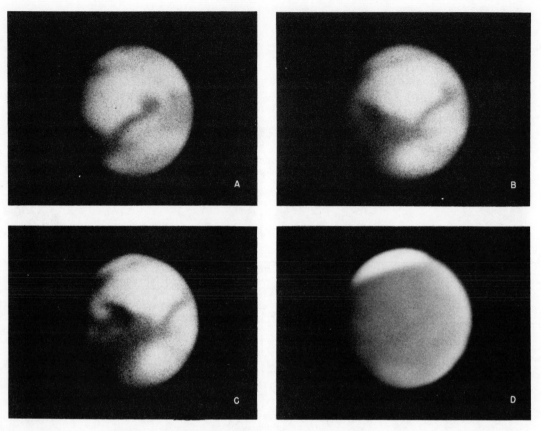

Figure 6.6 **Mars, Four Views**
Views A, B, and C, taken in succession (in red light), show the planet's rotation; D, in blue light, shows the atmosphere and the southern polar cap. Photographs from the Hale Observatories.

*effect,* converting longer-wave radiation to heat and holding it. The temperature of the upper atmosphere, however, appears to be low, somewhat below the freezing point of water on the dark side. What conditions may be on the surface, or even how far down the surface is, we are not sure.

## 6.3 MARS—ITS CHARACTERISTICS AND MOTIONS

Mars (Figure 6.6) is the first of the superior planets, and the last of the inner, or terrestrial, planets. It is smaller than the earth, a little over 4100 miles in diameter, and rotates rapidly enough to be measurably oblate. Its mass is approximately 10 percent the mass of the earth. This, coupled with its smaller diameter, gives it a surface gravity equal to about 40 percent that of the earth and a velocity of escape of 3.2 miles per second, approximately half that of the earth. It has an atmosphere, but a tenuous one, and its albedo is 4 percent. It is a better reflector in the visual range of the spectrum than in the photographic range because of its red color, to which the eye is more sensitive than a photographic emulsion.

Mars rotates on its axis in 24 hours and 37 minutes, so the day on Mars is approximately the same length as it is on the earth. The axis of rotation is tilted 25° to the plane of the orbit; consequently, seasonal changes on Mars are similar to those on the earth.

The orbit of Mars, although not as elliptical as that of Mercury or Pluto, is considerably more elliptical than the orbit of the earth. Mars' distance from the sun averages about 141 million miles but can vary by about 30 million miles during the Martian year. Since its average distance from the sun is 1.52 AU's, it receives less than half as much heat and light per square mile at its surface than the earth. Hence its average temperature is less than that of the earth.

Kepler's laws of planetary motion tell us that a planet 1.5 AU's from the sun has a sidereal period slightly under two of our years. Observation indicates that the sidereal period of Mars is 1.88 years, or 687 of our days. Its mean synodic period is 780 days, but the ellipticity of its orbit causes considerable variation from this mean.

When Mars passes behind the sun, it is said to be in *conjunction* with the sun. When the earth lies between Mars and the sun, Mars is said to be in *opposition* (Figure 6.7). Naturally enough, Mars is most clearly observed when it is at or near opposition. Since its distance from the sun varies from about 130 million to about 160 million miles, not all oppositions are equally favorable for

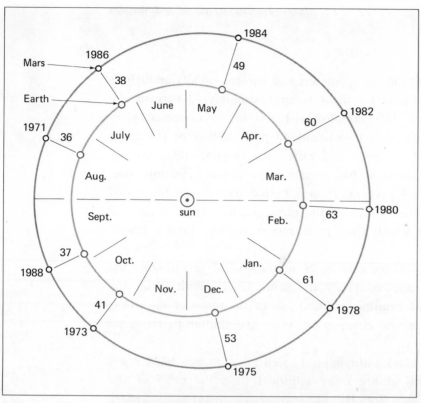

**Figure 6.7  Oppositions of Mars, 1971–1986**
Because of the eccentricity of Mars' orbit, oppositions that occur near its perihelion in late August are more favorable for telescopic observations from the earth than those at any other time. Oppositions occur about every two years and two months, as the earth gains a lap on Mars. At a favorable opposition, Mars is about 35 million miles from the earth; at its least favorable opposition, it is nearly twice that distance. In the drawing, the distances between the two planets are indicated in millions of miles. At most favorable oppositions Mars has an apparent angular diameter of approximately 25 seconds of arc.

observers on the earth. An opposition that occurs late in August is the most favorable because the earth is then between the sun and the perihelion of Mars' orbit. If an opposition occurs when Mars is near perihelion, the planet may be as close as 35 million miles from the earth and subtend an angle of about 25 seconds. At such times, it appears something less than $\frac{1}{60}$ the size of the full moon. If an opposition occurs in February, when the earth is in line with the aphelion of Mars' orbit, the distance from the earth to Mars is about 63 million miles.

**154**

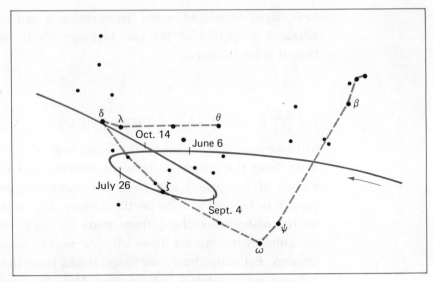

**Figure 6.8 The Favorable Opposition of Mars in August of 1971**
The diagram shows the path of Mars through the constellation of Capricorn
from May 27 to October 24, 1971. Mars was in opposition on August 10 and reached
its closest point to the earth (almost three-eighths of an AU) on August 12. North is
at the top, and the naked-eye stars in the vicinity are given their usual designations.

Because the synodic period of Mars is approximately two years
and two months, oppositions move up about two months in our
calendar each time we pass Mars. A favorable opposition (Figure
6.8) occurs about every 15 or 17 years. It is during such times
that most of the information regarding the planet obtained with
terrestrial telescopes has been gleaned. However, an observer looking at Mars for the first time, even during a favorable opposition,
is likely to be disappointed, so small does it appear to the eye
even with a powerful telescope.

Since Mars' axis is inclined approximately the same amount
as our own, it has seasons similar to those on the earth. However,
since the planet passes perihelion during the southern summer,
the southern summer is hotter and shorter than the northern
summer. Since the eccentricity of Mars' orbit is larger than the
eccentricity of the earth's orbit, and also because Mars does not

have large bodies of water to exercise a tempering effect, the seasonal inequality of the two hemispheres is more pronounced than it is on the earth.

## 6.4 THE MARTIAN ATMOSPHERE

Astronomers first deduced the existence of an atmosphere on Mars from the difference between infrared and ultraviolet photographs of the planet and from the occasional presence of what appear to be dust storms on the surface. For dust to move about on the planetary surface, there must be wind, and wind implies an atmosphere. As we have already noted, however, it is very tenuous. Estimates based on observations from the earth and from Mariner space vehicles indicate that Mars has less than 1 percent the atmosphere of the earth. Because the atmosphere is so thin, we are able to get a clear view of the planet's surface, comparable, but not quite equal, to the view we get of our own moon. The chief constituent of the Martian atmosphere is carbon dioxide. Its oxygen content is estimated by most authorities at less than 0.1 percent that of the earth's atmosphere, and by some authorities as absolutely negligible.

## 6.5 THE POLAR CAPS

Among the most interesting observable phenomena on Mars are the polar caps. As might be expected, their size varies with the season of the year. The southern cap disappears almost completely during the short, hot summer and extends over 3000 miles during the long, cold southern winter. The northern cap (Figure 6.9) does not get as big, nor does it disappear completely. The present consensus is that the caps are predominantly frozen carbon dioxide ("dry ice"). Between the regions of the cap, in the temperate and tropical

Figure 6.9 **The Northern Polar Cap of Mars**
The photograph, taken in 1972 by Mariner 9, shows the northern polar cap of Mars shrinking during the late Martian spring. Most of the material in the polar caps is thought to be dry ice, although some ordinary water ice is probably also present. NASA photograph.

zones, seasonal color changes take place. During the spring, the landscape appears to lose at least part of its reddish tinge, and a green or greenish-gray color overtakes some of the markings. Some observers have interpreted these changes as evidence of vegetation; others have suggested a chemical change as the cause. However, there is no proof that vegetation in any form exists on the planet. Certainly, the constitution of the atmosphere indicates that if vegetation or life of any kind does exist there, it must be quite different from that found on Earth.

## 6.6 SURFACE CONDITIONS ON MARS

Surface conditions on Mars are quite rugged by terrestrial standards. Because the atmosphere is tenuous, it does not provide the protection from the extremes of temperature provided by the at-

**157**

mosphere of the earth. Surface conditions on Mars have been described as being approximately like living on a terrestrial mountain twice as high as Mt. Everest. Temperatures probably exceed the freezing point of water in the tropics at noon, and perhaps reach as high as 25°C at the equator. On the other hand, the minimum temperature at the polar caps is usually estimated to be about −75°C. It is, however, difficult to relate these temperatures to earthly conditions because they are temperatures of the planet's surface rather than of the air (as temperatures are measured on earth) and may not be a true reflection of the relative comfort of the environment for man. Temperatures in the shade might cause a high degree of discomfort, even at the equator.

## 6.7 THE MARTIAN CANALS

In the past, some astronomers saw the seasonal color changes on the surface of Mars as substantiating the speculation that the red planet was the abode of life. During the opposition of 1877, the Italian astronomer G. V. Schiaparelli announced the discovery of what he called, in Italian, *canali* (Figure 6.10). This word, which means "channels," was mistranslated into English as "canals," giving rise to the idea that the channels had been artificially constructed. Late in the nineteenth century and in the early part of the twentieth century, speculation regarding the superintelligent inhabitants of the planet Mars who had presumably dug an irrigation system of great complexity was the order of the day, particularly in publications of the more sensational type. Some observers saw the canals as straight, fine lines that seemed to radiate like the spokes of a wheel from dark areas, appropriately named *oases*. Just about as many observers could not see the canals at all and denied their existence completely. The Mariner photographs of Mars show no evidence of the canals, and it has been suggested that they are perceptual effects—dots and other marks on the surface of

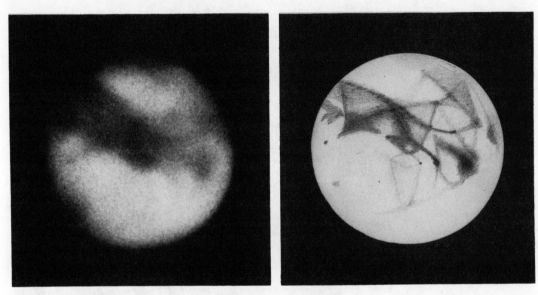

**Figure 6.10  The Martian Canals**
The photograph on the left and the drawing on the right differ because of the ability of the artist to wait until the terrestrial atmosphere is very steady, and the view of Mars exceptionally clear, to record details. The existence of the canals has not been confirmed by the Mariner photographs. Lick Observatory photographs.

Mars just beyond the range of definitive vision that are unconsciously connected and interpreted as lines.

## 6.8 THE MARINER SPACECRAFT

In the summer of 1965, the United States space vehicle Mariner 4 flew to within a few thousand miles of Mars, photographed the surface, and relayed the photographs back to earth. In the summer of 1969, Mariners 6 and 7 approached even closer to the surface and transmitted more photographs to earth. Finally, in the early 1970s, Mariner 9 achieved an orbit around the planet. This vehicle too took pictures.

The Mariner photographs were not transmitted as television

**159**

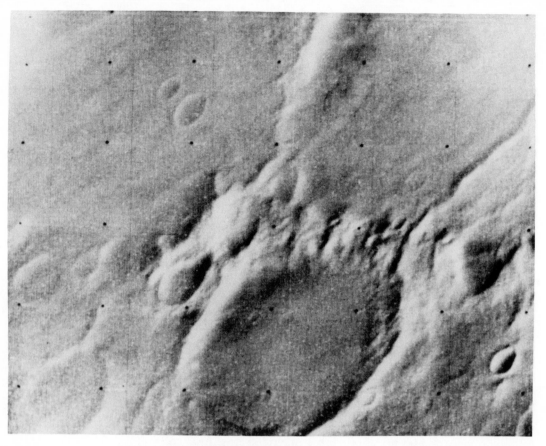

Figure 6.11 **The "Giant Footstep"**
Two adjacent Martian craters, foreshortened by oblique viewing. The area covered is about 85 by 200 miles. NASA photograph, Mariner 7.

pictures; the distance was simply too great. Instead, they were scanned by machine and the results of this scanning were taped in binary language. Mariner 4, for example, recorded 64 shades ranging from white to black. This record was then transmitted to earth and interpreted. The later, more sophisticated Mariner vehicles were able to handle considerably more shades of gray.

After examining the wealth of photographs received from these space probes, scientists concluded that there are no canals on the Martian surface. Indeed, a good bit of the surface of Mars seems to resemble very closely the surface of the moon (Figures 6.11 and 6.12). Other features of the red planet appear unlike anything

160

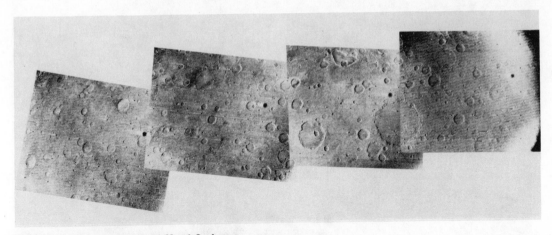

Figure 6.12 **Mosaic Photos of Mars' Surface**
Mariner 6 took these four photographs on July 30, 1969, at 84-second intervals. The area shown is about 450 by 2500 miles. NASA photographs.

found on either the moon or the earth. The Mariner 9 photographs, for example, revealed an extinct volcano much larger than any known on earth (Figures 6.13, 6.14, and 6.15) and a canyon wider and deeper than the Grand Canyon (Figure 6.16). The probes found no sign of life, no appreciable magnetic field, no free water, and an extremely tenuous atmosphere.

In addition to cameras, the Mariners carried *spectrometers*, which were used to obtain indirect measurements of atmospheric pressure and hence of the topography of the areas surveyed. These measurements indicate a range in altitude from highs of 3 to 4 miles to lows of 1 to 2 miles. One region in the southern hemisphere, Hellas, drops more than 3 miles from rim to center and rises again at the opposite rim. The two rims are about 1000 miles apart. The average atmospheric pressure is about 0.0053 of the terrestrial atmospheric pressure at sea level.

Remarkable photographs of Mars' two natural satellites were also obtained from Mariner 9. Both moons are extremely small. They were discovered in 1877 by Asaph Hall at the U.S. Naval Observatory. Called Phobos and Deimos (Fear and Panic), they

**161**

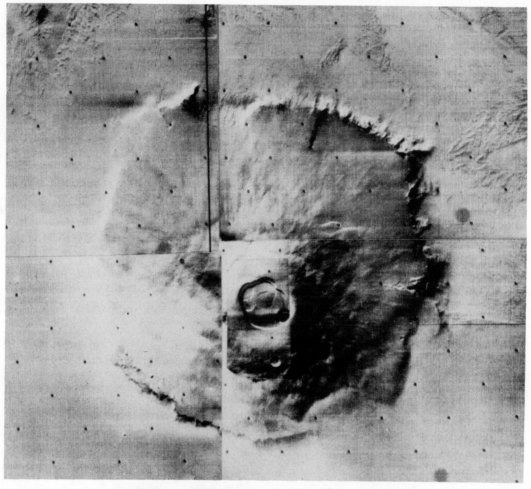

**Figure 6.13 The Volcano "Nix Olympica"**
This mosaic photograph by Mariner 9 reveals the largest volcano ever seen. Called Nix Olympica, it is over 300 miles across at the base. Steep cliffs drop from the mountain flanks to a surrounding great plane. NASA photograph.

are appropriate companions for the god of war. Phobos (Figure 6.17) is the larger of the two; Deimos is about half as big.

Before the Mariner program, the diameters of the two satellites could not be measured directly because they were too small. Their size was inferred from the amount of light they reflected, assuming an albedo approximately the same as that of the earth's moon. A Mariner 9 photograph of Phobos, taken at a distance of 3444 miles, shows it to be an irregularly shaped body about 10 by 14

162

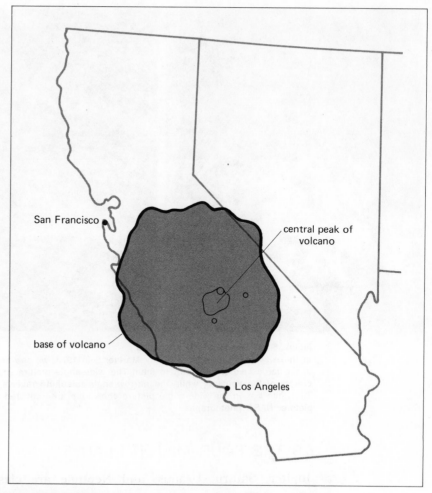

**Figure 6.14  Nix Olympica and California**
This scale drawing shows the relative size of Nix Olympica and the state of California. The volcano is so huge that it barely fits between Los Angeles and San Francisco.

miles in size, with a low albedo. Phobos is unique in the solar system in that it revolves around its primary faster than its primary rotates on its axis. Thus instead of rising in the east and setting in the west, it rises in the west and sets in the east. It makes a complete trip around the planet in about 7½ hours, rising three times each Martian day and moving eastward across the sky. To an observer on the Martian surface, of course, a satellite 10 miles wide would not appear very noticeable.

163

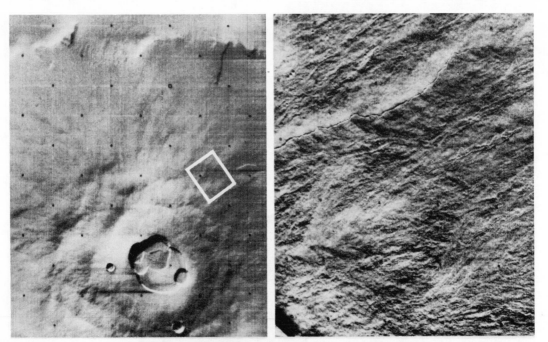

Figure 6.15 **Details of Nix Olympica**
In these two photographs taken by Mariner 9 in 1972, we see some of the details of the largest volcano known to man. The wide-angle picture shows the central craters of the volcano, while the narrow-angle telephoto reveals features on the side. The white box in the wide-angle picture shows the area covered in the narrow-angle picture. NASA photographs.

## 6.9 THE FOUR MAJOR PLANETS

Jupiter, Saturn, Uranus, and Neptune are classified as major planets. They are large, massive bodies lying between 5 and 30 AU's from the sun, with low average densities and heavy, opaque atmospheres. Of the 32 known satellites in the solar system, 29 revolve about the major planets.

## 6.10 THE MOTIONS OF JUPITER

The sidereal period of Jupiter is easily observed. It turns out to be 11.86 years. Its distance from the sun, according to Kepler's law, is 5.2 AU's. The earth gains a lap on its slow-moving big sister every 399 days, the planet's synodic period. The apparent motion

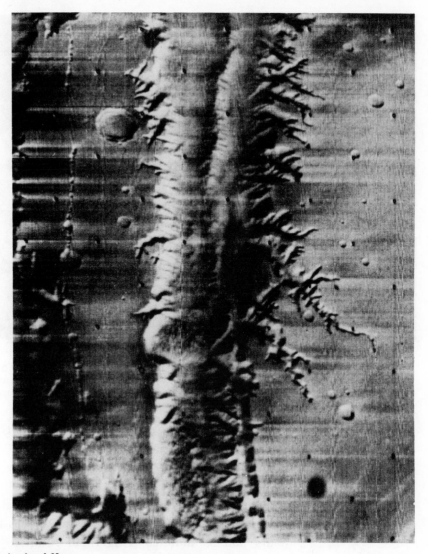

Figure 6.16 **The Canyonlands of Mars**
Mariner 9 revealed that the surface of Mars has vast canyons covering tens of
thousands of square miles. Some of the canyons are five times deeper than the Grand
Canyon in Arizona. The canyon shown in this view of Mars is located in Tithonius
Lacus, 300 miles south of the equator. The area covered in the photograph is
235 by 300 miles. NASA photograph.

of Jupiter is a stately progression eastward among the constellations. It travels once around the sky in about 12 years; retrograde
motions occur at intervals of about 13 months.

Since the inclination of its orbital plane is only 1.3°, Jupiter

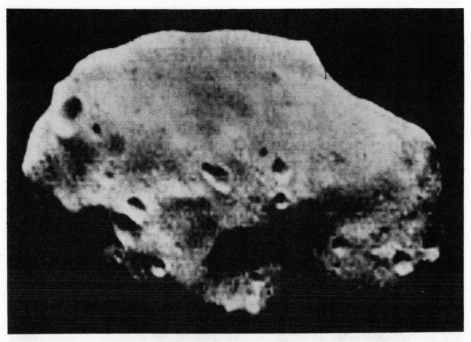

**Figure 6.17 Phobos, Satellite of Mars**
This remarkable photograph of Phobos was taken by Mariner 9 late in 1971. The
distance from the spacecraft was 5540 miles. Clearly, the surface of this small
satellite is covered with craters similar to those on our own moon. NASA photograph.

is never far from the ecliptic. Since its orbit is moderately eccentric, the difference in its distance from the sun at perihelion
and aphelion, about 47 million miles, is only 1 percent of the
mean distance.

The rotation period of the visible surface of Jupiter is not
the same in all latitudes, ranging from 9 hours and 50 minutes to
9 hours and 55 minutes. This rapid rotation, coupled with the
planet's large diameter, results in an equatorial velocity in excess
of 27,000 miles per hour. Jupiter's equator is inclined less than 4°
to the orbital plane; hence no seasonal effect is observed.

## 6.11 THE PHYSICAL CHARACTERISTICS OF JUPITER

Jupiter is the largest of the planets, with an equatorial diameter of
88,600 miles and a mass 318 times that of the earth. It contains

more matter than all the other known members of the solar system combined, excluding the sun. Moreover, the center of gravity of the sun-Jupiter system is outside the sun, making it unique in the solar system in this regard. However, the sun, with about 988 times the mass of Jupiter, is clearly the dominant member of the system.

Jupiter's mass is easily determined from the motions of its 12 known satellites. It also has a considerable effect on the orbits of comets and asteroids, perhaps even capturing some as satellites. Its average density, calculated from its mass and diameter, is only 1.34 times that of water, significantly lower than the density of a terrestrial planet. The density probably increases greatly as one descends from the visible, gaseous surface to the center, where it may exceed 30 times that of water. Jupiter, however, may be fluid to the center. It probably consists almost entirely of hydrogen and helium. Its fluid and gaseous character, coupled with its rapid rotation, gives it a very oblate shape.

The surface gravity of Jupiter is 2.4 times that of the earth, and its velocity of escape is 35.7 miles per second. At 5.2 AU's from the sun, the temperature is low, about −130°C. Jupiter has retained a thick atmosphere, keeping even a large quantity of hydrogen in its makeup. Spectral analysis of the upper layers of the atmosphere indicates that it consists largely of methane and ammonia, clouds of which arrange themselves in bright belts that alternate with darker stripes parallel to the equator, giving Jupiter its characteristic banded appearance (Figure 6.18). These clouds are also responsible for the planet's high albedo (over 40 percent).

An apparently permanent feature of Jupiter's atmosphere, the "great red spot," has been observed for over a century. It is an elongated oval area, dark in color and about 30,000 miles long. Its period of rotation is a little shorter than that of the surrounding atmosphere, and the difference accumulates to the equivalent of one rotation around Jupiter in about 10 years.

Jupiter's magnetic field is strong, perhaps ten times that of the

Figure 6.18 **Jupiter**
This photograph shows the characteristic banded appearance of the planet and the famous "red spot." Photograph from the Hale Observatories.

earth, and there is evidence of regions around Jupiter similar to the Van Allen belts around the earth. The problem of radiation from Jupiter is a complicated one. In addition to reflecting sunlight, Jupiter is one of the brightest objects in the sky in certain regions of the radio spectrum. Since 1955, emissions at a wavelength in the region of 10 centimeters have been measured. Recent measurements indicate that Jupiter is radiating as much as *2.5 times* the total energy attributable to reflected sunlight. If this figure is correct, some energy source must exist within Jupiter itself. For this radiation to be attributable to *thermal* (heat) sources would require temperatures between 40°C and 370°C. This is much greater than the observed temperature of −130°C; hence something other than heat (*nonthermal* sources) must be responsible.

168

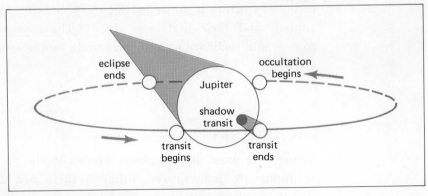

**Figure 6.19 The Motions of a Jovian Satellite as Seen from the Earth**
Jupiter's satellites may be observed even with a small telescope, although the size
of the satellite in comparison with Jupiter is greatly exaggerated in this drawing.
Predictions of the motions of the four Galilean satellites are published regularly in the
*American Ephemeris and Nautical Almanac.*

## 6.12 THE JOVIAN SATELLITES

Jupiter has 12 known satellites, more than any other planet. The
four largest, which range in diameter from 1800 to 3050 miles, were
discovered by Galileo in 1610. They can easily be seen with a small
telescope, and their daily positions with respect to Jupiter appear
in various almanacs and other publications. Since their orbits lie
near the plane of Jupiter's equator and hence close to the ecliptic,
eclipses of satellites by the planet (and transits) can be observed
from the earth (Figure 6.19). The innermost of the four large
satellites, Io, is particularly interesting. It is about as far from
Jupiter as our moon is from the earth, but it revolves around
Jupiter in 1 day and 18 hours instead of the 27.3 days it takes the
moon to go around the earth. This higher speed in an orbit of
approximately the same size is a direct result of the larger mass
of Jupiter, which we can compute by comparing the two orbits and
the two periods. Inside the orbit of Io, closer to Jupiter, is a
small satellite discovered by E. E. Barnard in 1892. Outside these
inner five, there are seven more. The inner three of these seven

**169**

revolve in orbits inclined between 25° and 30° to the equator of Jupiter, and their motion is direct. The four outer ones, with comparable inclinations and retrograde motions, may be captured asteroids.

## 6.13 SATURN

Saturn, the most distant planet known to the ancients, is similar to Jupiter in many ways, although there are differences. It is slightly smaller, with an equatorial diameter of about 74,000 miles, and more oblate. Its mass is 95 times that of the earth, but less than a third that of Jupiter. It is about 65 percent hydrogen and has a surprisingly low average density, 0.7 that of water. The velocity of escape at the visible surface is 23 miles per second, enough for it to have retained the hydrogen that existed in the original mass from which the planet was formed. The surface gravity is only slightly more than the surface gravity on earth, for although Saturn is nearly 100 times as massive as the earth, it has nearly ten times the radius. The increased attraction of the greater mass of Saturn is counteracted by its larger radius, which makes the surface farther from the center of the planet. Saturn rotates in a little over 10.6 hours near its poles, and in about 10.3 hours at the equator. The equator is tilted 27° to the orbit of the planet's orbital plane.

Saturn moves about the sun in a leisurely fashion at a distance of about 9.6 AU's, taking about 29.65 years to make the trip. It takes the earth only a year and 13 days to gain a lap on Saturn. When Saturn is at conjunction with the sun, it is nearly a billion miles away from us. Its orbit is nearly circular, with an eccentricity of 0.056.

Saturn receives only about 1 percent as much solar energy per square mile as the earth. Its atmosphere resembles that of Jupiter, although because of its lower temperature, less ammonia is evident

Figure 6.20 **Saturn**
Cassini's division between the two bright rings can easily be seen in this photograph.
Ring C, the crepe ring, is not visible, nor is the suspected faint outer ring.
Photograph from the Hale Observatories.

in the atmosphere, most of it probably having been frozen out. The temperature appears to be about −150°C, and the planet has an albedo of slightly more than 40 percent.

Saturn has ten satellites, the innermost of which was discovered only in 1967. They revolve in orbits ranging from a little less than 100,000 miles from the center of the planet to over 8 million miles. The largest, Titan, is 3076 miles in diameter and is the only satellite in the solar system known to have an atmosphere.

The most unique feature of Saturn is its ring system, a phenomenon associated with no other of the sun's planets. Galileo was unable to explain the changing appearance of Saturn—the strange protuberances which appeared and disappeared. In 1655, Huygens identified these protuberances as rings. The ring system is about 171,000 miles in diameter. It is flat and thin, a maximum

**171**

Figure 6.21 **Saturn with Rings on Edge**
From a drawing by Barnard, December 12, 1907. Yerkes Observatory, University of Chicago.

of about 10 miles thick, and lies in the plane of the planet's equator. In 1675, Cassini discovered a division in the ring, visible through a medium-sized telescope (Figure 6.20). In 1850, Bond discovered what is called the *crepe ring*, inside the bright ring. The three principal rings are lettered, working inward, A, B, and C. A fourth ring may lie outside ring A, but if so, it is very faint. Other divisions have been discovered since Cassini, at harmonics of the periods of the satellites, particularly Mimas, which is fairly close to the rings and has a perturbing effect upon them.

The rings were originally assumed to be solid structures, but such structures would be torn apart by the tidal pull of the planet. The fact that the inner portions of the rings move faster than the outer portions indicates that they are comprised of individual particles. If they were solid, the outer edges would move faster than the inner sections. Astronomers today believe that the rings are composed of small particles, resembling meteorites and perhaps covered with frost or ice. Stars can be seen through the

172

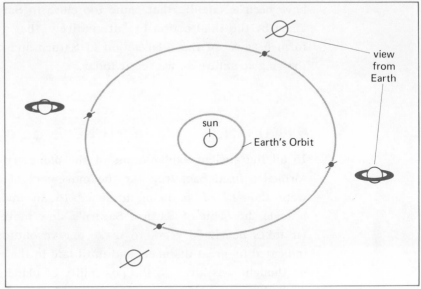

Figure 6.22 **Aspects of Saturn's Rings**
As Saturn revolves around the sun, the northern and southern sides of the ring
system, inclined about 27° to the plane of the orbit, are alternately presented to the
earth. Midway between the full "opening" of the rings, it is seen edgewise. The
entire sequence is completed in a sidereal period of Saturn, approximately 30 years.
The orbital motion of the earth makes it possible for us to pass through the plane
of the rings three times in rather rapid succession when they are edgewise.

ring, and it is estimated that the particles occupy about $\frac{1}{16}$ of the
volume of the system.

As Saturn goes around the sun, the rings as seen from the
earth are sometimes on edge (Figure 6.21) and at other times
tilted up to 28°. Approximately 7.5 years elapse between the edge-on,
or closed, phase and the face-on, or open, phase. Fifteen years
later, the rings again appear open, but we see them from the other
side (Figure 6.22).

It has been demonstrated that if a liquid satellite came within
2.4 radii of a primary of the same density, it would be torn apart
by the tidal action of the primary. The innermost satellite of
Saturn is outside this limit, but the rings lie entirely within it. It
has been suggested, therefore, that the material in the rings may

have been a satellite that came too close to Saturn and was torn apart by the tidal action or, alternatively, that a satellite failed to form because of the tidal action of Saturn and the particles filled the ring structure as we see it today.

## 6.14 URANUS AND NEPTUNE

In all the ancient explanations of the planetary system, the stars formed a fixed backdrop for the movement of the planets. They were thought of as being attached to an invisible sphere just outside the orbit of Saturn. Saturn's slow movement in its orbit (it takes nearly 30 years to make one revolution about the sun) indicated its great distance, and until late in the eighteenth century no thought was given to the possibility of planets beyond Saturn's orbit. Yet in the 150 years that followed, three more planets were identified. The story of their discovery is a fascinating one.

In 1781, William Herschel discovered the planet we call Uranus (Figure 6.23). Uranus is visible to the unaided eye if one knows exactly where to look, but since it is a faint object near the threshold of unaided vision, it was overlooked for centuries. Even after the invention of the telescope, when it was seen (as early as 1690), it was repeatedly plotted as a star. In 1781, William Herschel, inspecting an area of the sky along the ecliptic, perceived a small greenish dot that did not look like a star. Watching it from night to night and month to month, he saw it move and realized that it was a planet. He named the planet after King George III, but this appellation was considered inappropriate for a celestial body. Some astronomers referred to it for a while as Herschel's star, before the name Uranus was finally adopted.

Since Uranus' position had been plotted for decades before its identification as a planet, reliable data were available for calculating its orbit. When astronomers compared its actual path with the predicted orbit, however, they noticed various discrepancies. By the

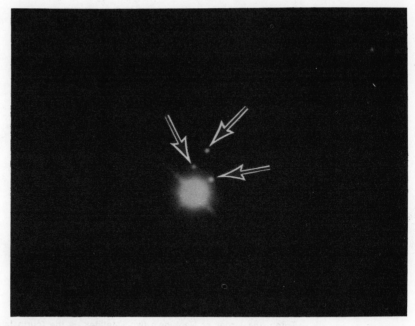

**Figure 6.23 Uranus and Three of Its Satellites**
The plane of the equator and the orbits of the satellites is nearly perpendicular to
Uranus' orbit. Lick Observatory photograph.

early 1840s, it had become evident that another unknown planet, not
accounted for, was disturbing the orbit of Uranus.

In England and in France, two astronomers, John Adams and
Urbain Leverrier, were working independently on this problem. Almost simultaneously, they sent their calculations to observatories.
Leverrier's figures went to Berlin and Adams' to Greenwich. For
some reason, Adams' calculations appear not to have been taken
seriously. In the Berlin Observatory, however, astronomers using
Leverrier's predictions discovered Neptune within 30 minutes of beginning their search. This was in 1846. Neptune (Figure 6.24) is a
faint object, some ten times fainter than the faintest object visible to
the unaided eye. Given the size of the telescopes available in 1846,
its discovery was no mean accomplishment.

The great distance of the outer planets from the sun and from

**175**

Figure 6.24 **Neptune and One of Its Satellites**
Triton, the larger of Neptune's two satellites, is the fourth largest satellite in the solar system. Its other satellite, Nereid, is extremely faint. Lick Observatory photograph.

the earth makes our information about them somewhat fragmentary, and the probable error of the estimates of their mass and dimensions is rather high. Since Uranus and Neptune are similar in size and constitution, we shall consider them together.

Although measuring their diameters is quite difficult, all indications are that the two planets are about the same size. Uranus is estimated to be about 30,000 miles in diameter and Neptune about 28,000. Uranus, despite its larger diameter, appears to have a smaller mass, about 14.5 times the mass of the earth. Neptune has 17 times the mass of the earth. Obviously, the density of Uranus is smaller also, about 1.5 times that of water, compared with 2.3 times that of water for Neptune. The velocity of escape on the two planets is quite similar, between 14 and 16 miles per second.

Since these planets have about 4 times the diameter of the earth and approximately 16 times the mass, their surface gravity is close to that of the earth. It is somewhat less, of course, on Uranus than on the more massive Neptune. Measures of oblateness are inconclusive.

Like Saturn and Jupiter, these two major planets rotate rapidly, Uranus in about 10.7 hours and Neptune in about 15.5 hours. Since no markings are clearly visible on either planet, their rotation must be determined from the Doppler shift in spectroscopic observations of their limbs as they move toward or away from the earth. Whereas the inclination of the axis of Neptune is moderate, about 29°, that of Uranus is unique. Its axis of rotation lies almost in its orbital plane; indeed, if we wish to regard Uranus as rotating directly, we must assume the tilt of the axis to be 98°, slightly beyond the planet's orbital plane. It is more logical to believe that the inclination is 82° and the rotation is from east to west.

Uranus, which is slightly over 19 AU's from the sun, revolves about it in 84 years. Its location seems to reinforce Bode's law, which predicts a planet beyond Saturn at 19.6 AU's. Bode's law does not, however, predict the position of Neptune, at 30 AU's. Neptune moves about the sun once in 165 years, and has not completed a revolution since its discovery. The synodic period for both planets is just a few days over a year. They move so slowly among the stars that the earth has only to move a few degrees after completing one revolution about the sun in order to gain a lap on them. Spectroscopic observations indicate atmospheres composed primarily of methane and hydrogen on both planets. They have high albedos, reflecting half and two-thirds of the light they receive. The approximate temperature on Uranus is −185°C; that on Neptune is even colder, −210°C.

Uranus has five known satellites and Neptune two. Uranus' five satellites have orbits in its equatorial plane, highly inclined to the ecliptic. The orbits range in size from 80,000 miles to 364,000 in radius. The diameter of each of the five satellites is less than 1000 miles. Neptune has one satellite, Triton, which is larger than our moon and could, theoretically, possess an atmosphere. Its smaller satellite, Nereid, has a very eccentric orbit and may be a captured asteroid. Its distance from Neptune varies by a factor of six.

## 6.15 PLUTO

One of the principal activities of the Lowell Observatory in Flagstaff, Arizona, is a study of the planets, and for many years a search had been conducted for what was called at the time a trans-Neptunian planet. Percival Lowell, the founder of the observatory, died before the planet was discovered, in 1930, by Clyde Tombaugh. It was named Pluto, however, and the symbol for the planet is composed of Lowell's initials, the letters PL.

So far as we know, Pluto (Figure 6.25) is the outermost planet of the solar system. If there are other planets beyond it, then they must be very faint indeed. One fortunate circumstance in the discovery of Pluto should be noted. Pluto's orbit is highly inclined, about 17°, to the ecliptic, and it is within the zodiac only near the nodes. Searches for planets are conducted along the ecliptic, and in 1930 Pluto was near the ecliptic, not far from one of its nodes.

Pluto is so far from the sun, 39 AU's, that direct observations of its physical characteristics are extremely difficult. It seems to be about the size of Mercury or Mars, with a mass between 0.6 and 0.8 that of the earth. For these figures to be correct, however, the planet would have to be abnormally dense. Since Pluto possesses no satellite, the figure for the mass is derived from its effect on the orbits of Neptune and Uranus. The great uncertainty about its size and mass make it impossible to say anything about the surface gravity. Since the visible disk is tiny and bears no markings, the rotation of the planet is also uncertain. Photoelectric observations of the variations in Pluto's brightness indicate that it may rotate in a period of about 6.4 days, turning alternately brighter and darker faces toward us. This seems to be the only explanation for the change in its brightness.

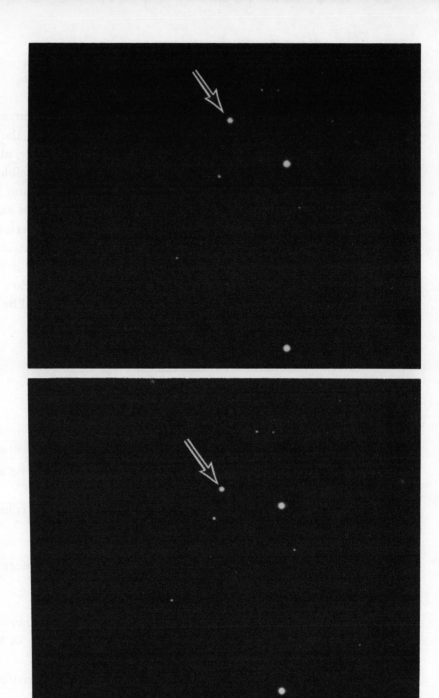

Figure 6.25 **Pluto**
These two photographs, taken with a 200-inch telescope, show the movement of the
planet in 24 hours. Photographs from the Hale Observatories.

The orbit of Pluto, in addition to being the largest of any known planet, is also the most high inclined (17°) and the most eccentric. At times, Pluto is closer to the sun than Neptune, although the high inclination of its orbit makes a collision impossible in the foreseeable future. Pluto is 100 times as far from the sun as Mercury and, in accordance with Kepler's law, takes 1000 times as long to revolve. Its sidereal period is 248 years, and its synodic period is slightly over one solar year. It probably has no atmosphere, although there is as yet no evidence on this point. Since its size is uncertain, its albedo is unknown. The temperature must be below −210°C. So far as we know, any variation in temperature would be due to the variable distance of Pluto from the sun.

# QUESTIONS

(1) At what time of the year can Mercury be seen most easily in the evening? Why? (Section 6.1)

(2) During which months of the year do transits of Mercury occur? About how often do they occur? (Section 6.1)

(3) Describe the atmosphere of Venus. (Section 6.2)

(4) Why are some oppositions of Mars more favorable than others for observing the Martian surface? (Section 6.3)

(5) Describe the atmosphere of Mars. (Section 6.4)

(6) What was the function of the Mariner series of space probes? What did they reveal about the surface of Mars?

(Section 6.8)

(7) Describe the basic physical characteristics of the planet Jupiter. (Section 6.11)

(8) How many moons has Jupiter? Which is the largest?

(Section 6.12)

(9) Describe the rings of Saturn.                    (Section 6.13)
(10) How were Uranus and Neptune discovered?
                                                     (Section 6.14)
(11) How long does a "year" last on Pluto?      (Section 6.15)

# CHAPTER 7
# OTHER MEMBERS OF THE SOLAR SYSTEM

In addition to the nine major planets and their 32 satellites, there are many other minor members of the solar system revolving about the sun. We shall consider in this chapter the asteroids, or minor planets, the meteoroids, and finally the comets. While the total mass of these other members of the system is insignificant, their variety of forms and characteristics makes them interesting in themselves. They also provide possible clues to the origin of the solar system. To slight them would be to leave our story of the sun's family only partly told.

## 7.1 THE MINOR PLANETS

As we saw in Section 4.13, Bode's law is a sequence of numbers that curiously parallels the sequence of distances of the major planets from the sun expressed in AU's. At a distance of 2.8 AU's, there is apparently a gap in the solar system, between Mars and Jupiter. Astronomers knew of no planet at a distance of 2.8 AU's until the night of January 1, 1801. On that evening, the Sicilian astronomer Giuseppe Piazzi discovered a small object which, from night to night, appeared to move among the stars. At first he thought it was a comet, but after observing it for a while, he became convinced that it was moving in an orbit between Mars and Jupiter. Before enough observations could be obtained to compute a definitive orbit of this small object, it approached too near the sun to be visible from the earth. Fortunately, the great mathematician Karl Friedrich Gauss had invented a new method of calculating orbits, which he applied to Piazzi's observations. On the last night of the year 1801, Ceres, the first asteroid, was found in the position predicted by his calculations.

In the early years of the nineteenth century, observations in astronomy were made by looking through the eyepiece of a telescope. Searching for asteroids in this way was singularly unprofitable with the telescopes then in existence. From 1800 to 1810, a total of four were found. These were the bright ones; a fifth was not seen until 1845. As telescopes became bigger, more asteroids were discovered. With the introduction of photography to astronomy at the end of the nineteenth century, the number of known asteroids greatly increased.

One way of finding an asteroid is to take a fairly long-exposure photograph of a star field along or near the ecliptic. If there is an asteroid in the field while the exposure is being made, its image will move. The stars will appear as dots in the picture, and the asteroid will be a small, but noticeable, streak. If very faint asteroids are suspected to be in a certain region and the photographer has some idea of the direction of their motion, he can move his telescope slightly to trail a particular asteroid; the stars will then be little streaks, and, hopefully, the asteroid will be a round dot. Fainter asteroids can be detected in this way, since their light is concentrated in one point. With a large telescope, the image of an asteroid may be recorded using a shorter exposure. A suspected asteroid may also be detected by comparing photographs taken at appropriate intervals and noting whether a particular dot has changed its position with respect to the stars.

There are literally tens of thousands of asteroids bright enough to be seen with a 200-inch telescope. In the *National Geographic Society-Palomar Observatory Sky Survey*, thousands of asteroid trails were recorded as pictures of the stars throughout the sky were taken.

## 7.2 THE ORBITS OF THE ASTEROIDS

An asteroid is not regarded as confirmed, or "in captivity," until at least a preliminary orbit has been calculated and the body has been rerecorded at a subsequent opposition. The orbits of the asteroids show some regularity. Like the major planets, they revolve from west to east. The inclinations of known asteroid orbits to the ecliptic are for the most part small, although they are, on the average, larger than the inclinations of the orbits of the major planets. (In one case, the inclination exceeds 50°.) Although most of the asteroids are in a belt about 0.5 AU on either side of the 2.8-AU orbit predicted for Bode's "missing" planet, not all of them are in this region. At perihelion, Icarus, which has a mean distance from the sun about equal to the earth's distance but whose orbit is very eccentric, is inside the orbit of Mercury. Figure 7.1 shows another extreme example, Hidalgo, which at aphelion is outside Jupiter's orbit in the vicinity of Saturn.

Several asteroids come within 3 to 6 million miles of the earth. When this happens, it becomes possible to refine the length of the astronomical unit by measurements of their parallax. By analyzing

Figure 7.1 **Eros**
This photograph shows the trail of the asteroid as seen through a 40-inch refracting telescope on February 16, 1931. Yerkes Observatory photograph, University of Chicago.

the observed positions of the nearby asteroid from different points on the surface of the earth, we can obtain the direct parallax by the usual method of triangulation. Since the positions of the asteroid can be computed from its orbit, this gives us an immediate measure of one distance within the solar system in terms of statute miles (or kilometers) and hence establishes a scale for the solar system. The first exhaustive attempt in this area was carried out using observations of the asteroid Eros when it approached the earth in 1931 (Figure 7.2).

One group of asteroids is of particular interest. The eminent French mathematician Joseph Lagrange predicted that, ahead of and behind a massive planet, 60° from the planet itself, there would be

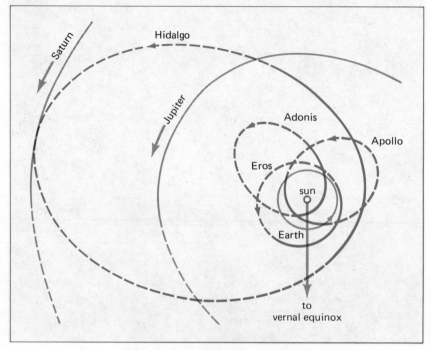

**Figure 7.2  The Orbits of Four Asteroids**
The asteroids shown are not typical. Eros is among those which come closest to the earth. Hidalgo goes farther from the sun than any other known asteroid.

two points (since named the Lagrangian points) at which a body would tend to stabilize in orbit. An examination of these two points ahead of and behind Jupiter has revealed 14 asteroids. Called the *Trojan group* and named for the warriors of Greece and Troy, they oscillate in two clusters in the immediate vicinity of the two Lagrangian points (Figure 7.3). Possibly they have been captured by the gravitational action of Jupiter and will at some future time escape from it.

Jupiter also has another effect on the orbits of the asteroids. Just as the satellite Mimas produces gaps in the rings surrounding the planet Saturn (Section 6.13), so too Jupiter produces gaps in the asteroid orbits. Basically, these gaps, called the *Kirkwood gaps* after

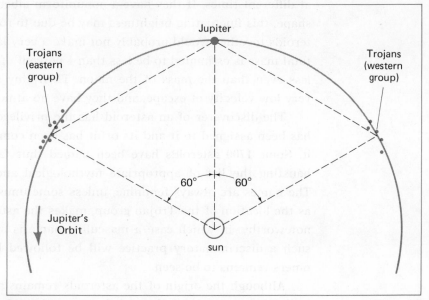

**Figure 7.3  The Trojan Asteroids**
These asteroids precede and follow Jupiter in its orbit, 60° ahead of and behind the planet. The two points about which they cluster are called the Lagrangian points and are produced by the gravitational force of Jupiter.

their discoverer, appear where the period of revolution of the asteroid is a simple harmonic of Jupiter's period.

## 7.3 PHYSICAL FEATURES OF THE ASTEROIDS

All the asteroids are small. Perhaps 10 or a dozen are as big as 100 miles in diameter, and most of the rest are even smaller. Many are only a few miles in diameter, and what the lower limit is we cannot tell. Ceres, the first asteroid to be discovered, is about 500 miles in diameter. It is difficult to measure the disk of even the largest of such small objects. If we assume an average albedo of about 10 percent, however, then we can use the light reflected by an asteroid to estimate its size. Some asteroids reflect different amounts of light

187

at different times. If they have a nonuniform albedo or an irregular shape, this fluctuating brightness may be due to rotation. All the asteroids together would probably not make a very large object. Their total mass is estimated to be less than 1 percent of the earth's mass, less even than the mass of the moon. Their tiny size gives them a very low velocity of escape, and they have no atmosphere.

The discoverer of an asteroid has the privilege, after a number has been assigned to it and its orbit has been computed, of naming it. Some 1700 asteroids have been named thus far, practically exhausting the list of appropriate mythological and literary names. The names are always feminine, unless some unusual feature, such as the location of the Trojan group, makes the asteroid particularly noteworthy, in which case a masculine name is assigned. Whether such a discriminatory practice will be followed by future astronomers remains to be seen.

Although the origin of the asteroids remains an open question, it has been suggested that in the formation of the solar system either of two things may have happened. A planet may have formed and been disrupted by the tidal action of the great mass of the planet Jupiter, and the pieces gone into orbits of their own. Or, a planet may have failed to form and the pieces gone into orbits approximating the orbit the planet would have followed. What evidence there is does not particularly favor one theory over the other. The asteroids, particularly the small ones, are generally not spherical, and they could well have originated in either of these two ways.

## 7.4 METEORS

Almost everyone has observed "shooting stars" or "falling stars" at one time or another. These are *meteors*; the streak of light observed is a *meteor trail* (Figure 7.4). Most meteor trails are produced high in the earth's atmosphere. Before a meteor reaches the earth's at-

Figure 7.4 **A Meteor Trail**
In this picture, we see a meteor trail in the constellation of Scutum. Yerkes
Obseratory photograph, University of Chicago

mosphere it is, by agreement, called a *meteoroid*. If it survives the
trip through the atmosphere and strikes the earth's surface, it is
called a *meteorite*. The study of meteorites is called *meteoritics*.

Meteroids are particles of stone or metal revolving around the
sun. When they strike the earth's atmosphere and burn up, they become visible. Since they have high velocities with respect to the
earth (from 8 to 45 miles per second), the amount of heat energy
they generate is considerable. It is this heat energy released in our
atmosphere that causes meteor trails. Meteoroids striking the atmosphere of the earth become visible at an altitude of about 80 miles,
and the majority are consumed rather quickly. Extremely bright
meteors have special names. A *fireball* is a meteor that is bright

189

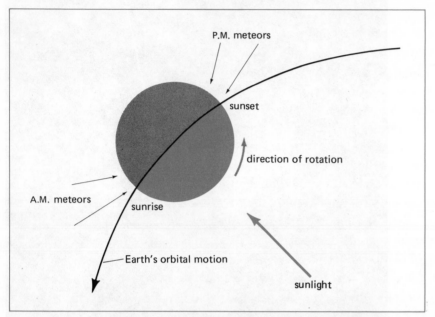

Figure 7.5 **Morning and Evening Meteors Compared**
In the evening, after sunset, the visible meteoroids are overtaking the earth; hence their velocity is equal to their own velocity in space *minus* the earth's velocity of about 18 miles per second. In the morning, before sunrise, the visible meteoroids are meeting the earth head on; their relative velocity is the sum, rather than the difference, of their own and the earth's velocities.

enough to cast a shadow. If a meteor explodes with an audible noise, it is called a *bolide*.

The number of meteroids must be in the trillions or more. It has been estimated that approximately 100 million or more meteor trails bright enough to be seen with the unaided eye against a dark sky occur all over the earth during a single day.* Since half of them appear in daylight, and because most of the earth's surface is uninhabited, the majority are unobserved.

Although meteors can be seen throughout the night if the sky is clear, they seem more numerous and brighter as the night pro-

*This figure is an extrapolation of the number of observations made by competent observers.

190

gresses. This is because just after sunset we are at the rear of the earth as it moves in its orbit. As Figure 7.5 shows, meteoroids entering our atmosphere at this time overtake the earth, and the velocity with which they strike the atmosphere is equal to their own velocity minus the orbital velocity of the earth. The resultant velocity can be as small as about 8 miles per second. This velocity is increased slightly by the gravitational attraction of the earth. As the earth turns, we move through 180° with respect to the sun, so that just before sunrise we are at the front of the earth as it speeds along in its orbit. At this time the velocity of the meteoroid with respect to the earth is equal to its own velocity *plus* the earth's orbital velocity. The resulting speed can be as high as 45 miles a second. This situation is analogous to that of a car being driven on a freeway during a rainstorm. Many raindrops hit the front windshield, while only a few drops fall on the rear window. During the early morning hours, meteors seem more numerous because of their increased velocity, which means that smaller masses produce visible trails. A meteoroid that might not be visible in the evening may produce a visible trail in the morning.

Sometimes a meteor is bright enough to produce, along its trail, what is called a *train*. This is a cylinder of ionized atmospheric gases and meteoric dust that may remain visible for a considerable time after the meteor has been completely consumed.

Over the past decades, considerable research has been done on the triangulation of meteor trails. The principle is the same as in other types of triangulation. A baseline 20 or 30 miles long is established, and synchronized cameras at either end are used to take simultaneous photographs of the sky (Figure 7.6). The shutters are opened and left open to pick up any meteor bright enough to register. In front of each lens is a rotating shutter, which alternately covers and uncovers it. If a meteor trail registers on the film, it will be interpreted at a known rate by the shutter. Thus instead of a streak, the picture will contain dashes interrupted by black areas. The length of the dashes enables the observer to time the meteor.

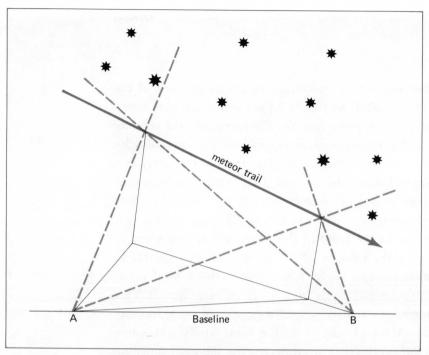

Figure 7.6 **Triangulating a Meteor Trail**
Synchronized cameras at points A and B take simultaneous photographs of any meteor
that happens to come into view while their shutters are open. This picture is
projected against the background of the stars, and the difference in the view from
the two points is used to compute the position of the beginning and end of the trail,
and consequently the orbit of the meteor. The meteor is timed by synchronized
rotating shutters, one in front of each camera. Counting the number of interruptions
of known duration gives astronomers its flight time, or speed.

Knowing its velocity, we can, by triangulation of the beginning and
end of the trail, calculate its orbit.

This technique, and others, has given us some knowledge of the
orbits of meteoroids. If meteoroids characteristically entered the atmosphere of the earth at velocities in excess of the velocity of escape from the solar system at the earth's distance from the sun, we
could conclude that they came from outer space and were not members of the solar system. However, the vast majority of observed
meteors have elliptical orbits.

Radar is also used in studying meteor trails. The ionized material in the trail reflects radar pulses and provides information about
the motion of the meteor. This data too seems to support the conclusion that meteoroids are members of the solar system. Apparently,

Figure 7.7 **The Zodiacal Light**
This photograph was taken from Haleakala, Hawaii, by P. Hutchinson, University of Hawaii, January 1967.

the creation or origin of the solar system was not an orderly, neat process in which a cleanup followed the main event. Considerable debris appears to have been left in the spaces between the planets, and the meteoroids may be the result of this leftover material smashing into the earth's atmosphere.

Small meteoroids, which can be as tiny as dust particles, are related to two observable phenomena, the *gegenschein* (counterglow) and the *zodiacal light* (Figure 7.7). The gegenschein is a faint glow of light visible in a dark, cloudless sky directly opposite the sun. One explanation of this hazy, faint light is that it is an aggregation of small particles held by the earth's gravitational pull. Another explanation is that the earth itself may have a meteoric "tail," although there is no direct evidence to support this contention.

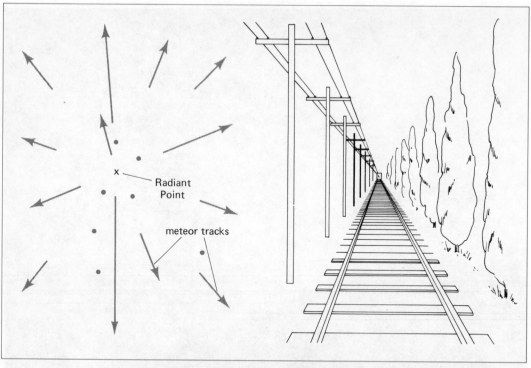

**Figure 7.8  The Radiant Point of a Meteor Shower**
If, during a meteor shower, an observer makes a record of each trail, the meteors
that are members of the shower appear to radiate like the spokes of a wheel from
a particular point in the sky, just as the tracks of a railroad appear to converge
in the distance. The meteoroids are traveling in paths that are nearly parallel; hence,
when they strike the earth's atmosphere, the effect of perspective makes them
appear to radiate from a point. The swarm or shower is usually named after the
constellation from which it appears to come, although the meteors are not physically
connected with the constellation or the stars in it.

The zodiacal light is a triangular wedge of light that points upward from the horizon in the west after sunset or in the east before
sunrise. Some think that this light is produced by reflections from
meteoric particles in the general plane of the solar system.

Chemical analyses of meteors can be carried out by allowing the
light of a bright trail to pass through a spectroscope. Since meteors
are basically unpredictable, this is a long and frustrating task, but
surprisingly good spectrograms of meteors in flight have been observed. The majority of the lines that appear are those of metals
and of nitrogen and hydrogen. The metals are the same as those
found in chemical analyses of meteorites (see Section 7.5).

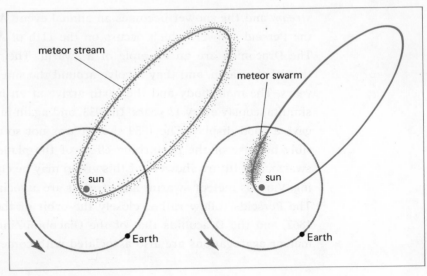

**Figure 7.9  Meteor Swarms and Streams in Orbit**
The appearance of a swarm of meteoroids produces a meteor shower. However, the
orbit in which the meteoroids travel must intersect the orbit of the earth at a time
when the earth is at the intersection point. If the meteors are more or less uniformly
distributed along the orbit, they form a stream, and meteors are seen each year. If
they are bunched up, a meteor shower will occur only when the swarm and the earth
pass the intersection point of the two orbits simultaneously.

Occasionally, meteors are more numerous than usual. If, when
this occurs, the meteor trails are plotted on a star map for the region, the trails generally appear to emanate from a certain point or
small region of the sky (Figure 7.8). This is called a *meteor shower*.
It occurs because the earth's orbit intersects the orbit of a swarm
or stream of meteoroids traveling together in the same path (Figure
7.9). The meteor trails appear to diverge from an infinite point,
much as the rails of a railroad track appear to converge in the distance. If the meteoroids are concentrated in one part of the orbit, the
aggregation is called a *swarm,* and meteor showers occur only when
the earth crosses the part of the meteoroid orbit where the swarm
happens to be. The following year, when the earth returns to the
intersection, the swarm may have moved on and no shower will
occur. If the swarm fills up the whole orbit, it forms a meteor

195

*stream* and the shower becomes an annual event. A good example is the Perseid shower, which occurs on the 11th of August each year. The Draconids are an example of a swarm. Their orbit intersects that of the earth, and they revolve around the sun in a period of $6\frac{1}{2}$ years. The main body and the earth arrive at an intersection point simultaneously every 13 years. In 1933, and again in 1946, this swarm gave a good display. The 1959 shower was not so conspicuous, possibly because of the perturbing effect of the planet Jupiter on the swarm, and future showers of this group may be equally disappointing. Various meteor swarms and streams are associated with comets. The Perseids follow rather closely the orbit of the great comet of 1862, and the Draconids that of the Giacobini-Zinner comet. Other meteor aggregations are also associated with cometary orbits.

## 7.5 METEORITES

Although the falling of "stones from heaven" has been recorded since the seventh century B.C., it is only in the past two centuries that scientists have generally accepted the idea that extraterrestrial material could actually be found on the surface of the earth. Meteorites vary in size, from very small ones to one of about 50 tons. The largest known meteorite lies near Grootfontein, in Southwest Africa. The largest meteorite in a museum is in the American Museum of Natural History in New York City. It was found in Greenland and weighs about 34 tons. The largest meteorite found in the United States is the Williamette meteorite, with a mass of about 14 tons. About 1600 separate falls have been cataloged. Reputedly, the largest number of individual falls is represented in the collection of the Chicago Museum of Natural History. Both the Smithsonian Institution in Washington, D.C., and the Griffith Observatory in Los Angeles, California, also have fine meteorite collections.

A number of spectacular meteorite falls have occurred. Two in Siberia in 1908 and 1947 did some destruction. Meteorites also

Figure 7.10 **The Barringer Crater**
This three-quarter-mile crater is near Winslow, Arizona. It was formed by the impact
of a meteorite, which was probably completely fractured on impact since tons
of meteoritic material have been picked up in the vicinity. Yerkes Observatory
photograph, University of Chicago.

struck a building in Illinois, a woman in Alabama, and a street in
the Ukraine. In addition, meteorite craters, some resembling lunar
craters, are found on the surface of the earth. The best known is the
Barringer crater, near Winslow, Arizona, which is about three-quarters of a mile wide and 600 feet deep (Figure 7.10). Tons of meteoric
iron were picked up within a few miles of the crater, and it is believed to have been formed fairly recently (that is, within the last
100,000 years or so). Attempts to locate the meteorite that formed
the crater by drilling have been unsuccessful. The meteorite itself,
of course, would be much smaller in diameter than the crater it
formed. Other craters suspected to be of meteoritic origin have been
found in Quebec, in Australia, and in other parts of the earth.

It seems probable that craters are formed by meteoroids which

**Figure 7.11 An Iron Meteorite**
A polished and etched slice of an iron meteorite shows its unique crystalline structure. This "Widmanstätten Pattern" identifies the iron as being of extraterrestrial origin. It was produced by the very slow cooling of the object at some time in its history. Dudley Observatory photograph.

overtake the earth and hence have a lower-than-average velocity. It seems unlikely that a head-on collision with a meteoroid would produce a crater, since the resultant high velocity would generate sufficient energy to burn up the object before it reached the earth's surface.

Physically and chemically, meteorites fall into three broad categories. There are *irons*, which are generally about 90 percent pure iron and 9 percent nickel, with a trace of other elements; there are *stones*, which have a characteristic structure; and there are *stony irons*, which range from irons with particles of stone to stones with flecks of iron. The irons can be recognized by a characteristic crystalline pattern in their structure (Figure 7.11). If the surface of an iron meteorite is polished and then etched with an acid, this characteristic crystalline pattern will appear. It is found in no other iron

Figure 7.12  **An Iron and Stone Meteorite**
A slice of a meteor found in Kiowa County, Kansas, shows its mixed composition.
Dudley Observatory photograph.

and identifies meteoric material. Only an expert, however, can identify a stone meteorite. Many pieces of slag and other odd-looking stone or metallic objects are brought to experts for identification each year. Sometimes they turn out to be meteorites; more frequently, they are simply terrestrial stones or other objects of terrestrial origin. Figure 7.12 shows an example of an iron and stone meteorite.

An examination of meteorites for radioactivity provides a means of determining the age of the material in much the same way that the radioactivity of rocks on the earth's surface allows us to date them. The average age of meteorites thus tested is approximately the same as the age of the earth, although many younger ones have been found.

In December of 1970, the National Aeronautics and Space Administration (NASA) announced the discovery of amino acids, the principal building blocks of living cells, on a meteorite that fell in Australia on September 28, 1969. The NASA release admitted that although terrestrial contamination of the meteorite was not entirely

Figure 7.13  **Tektites from the Philippine Islands**
Tektites may be the remains of material ejected from the earth's surface by meteorite impact, since they have compositions characteristic of the earth's surface. They appear to have been ejected through the atmosphere, to have cooled in space, and then to have returned to the earth's surface. Examples have been found at various locations on earth. A comparison of their chemical composition with that of lunar rocks indicates that tektites were never part of the moon. Dudley Observatory photograph.

ruled out, it was unlikely. This discovery suggests that these complex building blocks have been present from the time of the formation of the earth, since the age of the meteorite is about 4.5 billion years.

*Tektites* (Figure 7.13) are another class of objects believed to come from outer space. These small, glassy objects have been found in Australia and other places on the earth's surface, although they are not uniformly distributed. Some scientists have suggested that tektites are splash droplets from the moon which escaped after being formed by the impact of meteorites on the lunar surface. Others believe that the tektites were formed by meteorites striking the earth, progressed through the atmosphere, and fell back to the surface. Little certainty can be attached to these theories.

## 7.6 COMETS

Comets have been known from ancient times, and even today they are occasionally regarded with superstitious awe. It was first believed that comets were an atmospheric phenomenon, but to Kepler and other observers their lack of parallax indicated that they were farther away than the atmosphere of the earth could be supposed to extend. When a comet first comes within the range of a telescope, it appears as a small, hazy dot. Its appearance changes as it approaches and later recedes from the sun. Contrary to popular opinion, not all comets have tails. They do, however, have a *coma*, or head. Frequently, a small starlike point called a *nucleus* appears in the center of the coma, and it is then this point which is used to determine the position of the comet. The nucleus is small, probably not more than 50 miles in diameter. As a comet approaches the sun, it appears to move among the stars from night to night, generally growing brighter. It may or may not have a tail. After moving around the sun, it recedes once more to the more distant reaches of the solar system in an elliptical or parabolic orbit.

The most famous of all comets is Halley's comet (Figure 7.14). Edmund Halley was studying the orbit of the comet of 1682 when he noticed its surprising similarity to the orbits of two other bright comets, those of 1531 and 1607. Noting the common interval of approximately 75 years between these celestial phenomena, he predicted the return of the comet in 1758. Although he did not live to see his prediction come true, the comet was picked up on Christmas night in 1758 and passed perihelion in the spring of 1759. With one exception, returns at intervals of about 75 or 76 years have been recorded from 240 B.C. Halley's comet was named in his honor because he was the first to recognize that it was a periodic comet and hence a permanent member of the solar system. It was Halley's comet that was in the sky at the time of the Norman conquest of England in 1066. It last appeared in 1910 (Figure 7.15) and passed aphelion beyond the orbit of Neptune in 1948. It should

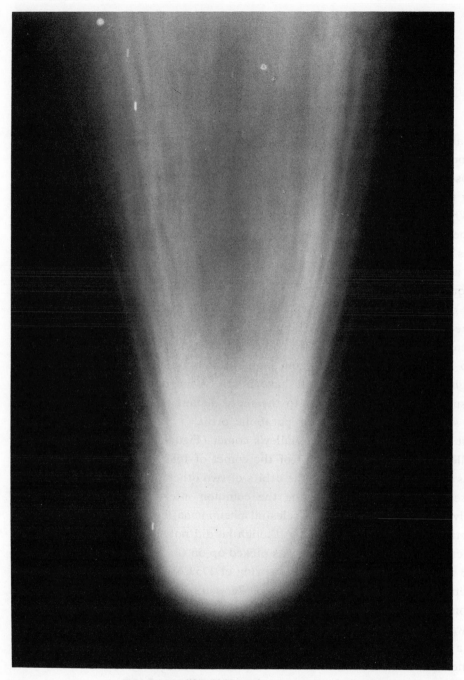

Figure 7.14 **Halley's Comet**
This picture, taken on May 8, 1910, with a 60-inch telescope, shows the head of
Halley's comet. Photograph from the Hale Observatories.

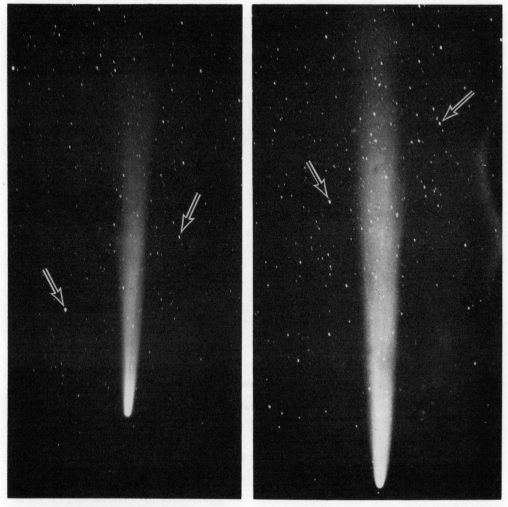

Figure 7.15 **The Movement of Halley's Comet**
The pair of stars indicated by arrows in the two photographs are the same. A comparison of the position of the comet show its apparent movement in 72 hours, from May 12 to May 15, 1910. Photographs from the Hale Observatories.

be back again in 1986, when the best view will be from the Southern Hemisphere.

Halley's comet is the only bright comet with a period of less than a century. Its motion is retrograde, and its elliptical orbit is extremely narrow and elongated (Figure 7.16). It comes inside the orbit of the earth at perihelion, and the earth passed through its tail in May 1910 with no observable effect. From its period of 75

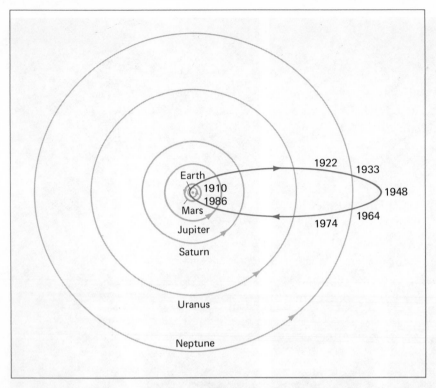

Figure 7.16  **The Orbit of Halley's Comet**
Halley's comet is the brightest comet known with a period of less than a century.
It last passed perihelion in 1910 and will return to view about 1986. It is now
between the orbits of Neptune and Uranus. The motion is retrograde.

years, one can easily calculate its mean distance of 17.8 AU's. Its
distance at perihelion is about 0.6 AU's.

The most distinctive characteristic of a comet is its orbit.
When a comet is discovered following an orbit that is the same
or very similar to the orbit of earlier comets that appeared at
regular intervals, it is identified as a *periodic* comet. Periodic
comets move about the sun in elliptical orbits. Most have somewhat
eccentric orbits, although in two cases the orbits approach a
circular form. Ortena's comet has a period of 8 years and revolves
entirely between the orbits of Mars and Saturn. The Schwassmann-
Wachmann comet has a period of 16 years, and its orbit ranges
between the orbits of Jupiter and Saturn. These are exceptions;
all other comets have more elongated elliptical orbits. Occasionally,
a comet appears to be traveling in a parabolic orbit, but since

comets are observed only when they are near the sun and a very long ellipse looks very much like a parabola, at least near the end, it is probable that either these are very elongated ellipses or the comet, in approaching the sun, has been accelerated by some of the larger planets into a parabolic orbit. Occasionally, a comet appears to be following a hyperbolic orbit near the sun, but again this is probably the result of planetary acceleration. So far as we know, comets are members of the solar system and travel in closed orbits.

Comets may be divided into two groups, according to the characteristics of their orbits. Those with nearly *parabolic orbits* have high inclinations to the ecliptic. Approximately half of these comets revolve from west to east; the others have a retrograde motion. Comets with definitely *elliptical orbits*, which we say are periodic, have moderate inclinations and, for the most part, direct (west-to-east) motion.

Of this latter group, about 40 or 45 are in what is called *Jupiter's family of comets* (Figure 7.17). This group of comets all have direct motion and reasonably small inclinations, and their aphelion points and one node are near Jupiter's orbit. Thus they may have been captured by Jupiter; that is, their orbits may at one time have been longer and more elliptical. If this is true, Jupiter's family should not be a permanent aggregation and may gain or lose members in the future.

Some comets follow the same path or nearly the same path near the sun as earlier, but different, comets. An example is the Ikeya-Seki comet of 1965. Near the sun, it was following essentially the same orbit as the great comet of 1882. Both comets were probably the result of the disruption of a former, much larger comet by the sun's tidal action. Near the sun, the returning parts follow the same or nearly the same path as the parent comet; but due to differences in the major axes of their highly eccentric elliptical orbits, they return at different times.

Most comets are discovered by accident, although at one time they were avidly sought by professional astronomers. This avidity

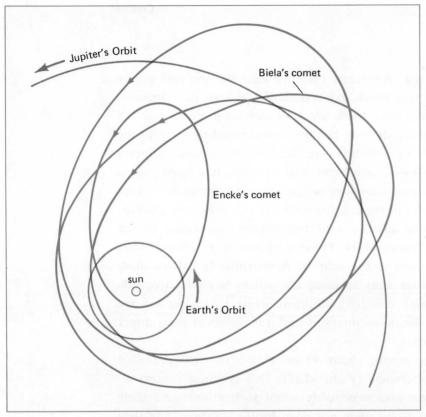

Figure 7.17  **Jupiter's Family of Comets**
The family of comets usually associated with Jupiter contains several dozen members, only a few of which are shown here. The aphelions and a node of their orbits are always near Jupiter's orbit. Probably, these comets were captured by the attraction of the massive planet. Membership in the family may not be permanent.

may have been encouraged by the awarding of medals and cash prizes for discovering comets, a practice no longer in vogue. Amateurs frequently discover comets, and several amateurs have discovered a number of them. The chief requisite is a small telescope, a great deal of patience, and a catalog of nebulae (to prevent confusion of these objects with suspected comets). Comets are most often found near the western horizon just after sunset and near the eastern horizon before sunrise. They are frequently discovered by accident on photographs taken for other purposes. *The National Geographic Society-Palomar Observatory Sky Survey* turned up a number of comets. They also have been discovered near the sun during total eclipses.

When a comet is discovered, it is given a provisional designation consisting of the year of discovery and a lower-case letter indicating its chronological order in a list of comets discovered that year. Thus the first comet discovered in the calendar year 1975 would be designated "1975a." After a comet's orbit has been computed, it is given a permanent designation, which consists of the year of its perihelion passage and a Roman numeral indicating its order among the comets passing perihelion during that year. 1975 II would be the second comet to pass perihelion in 1975. The year of discovery and the year of perihelion passage need not be the same. Bright comets are generally given the name of their discoverer or discoverers, or the name of a person who has done significant work on them. Halley's comet is an example of this latter category.

The outstanding physical feature of comets is their extremely low average density. They contain a very small amount of matter spread over a large volume. Estimates of the mass of comets range from 1 billionth to perhaps 1 trillionth of the mass of the earth. We know that comets have very little mass not by how they affect other bodies in the solar system, but by what they fail to do. When Brooks' comet, for instance, passed very close to Jupiter, its own orbit was greatly changed but it had no measurable effect on even the very smallest of Jupiter's satellites. Since comets are large, with heads 50,000 or 100,000 miles in diameter and tails 100 million miles long (or longer in extreme cases), the volume through which their low mass is spread is huge. This results in an extremely low average density. Figuratively speaking, one could take all the matter in the tail of a comet, pack it into a suitcase, and walk off with it. The density in the nucleus is about 1 millionth that of air at sea level, but even this is much greater than the density of the tail.

How can an object of such low density and such great size, containing such a small quantity of material, be so spectacular? Astronomers believe that comets consist of ices (frozen gas) and small dust particles. When a comet is far from the sun, the central portion is inactive. As it approaches the sun, the effect of the

**Figure 7.18  The Comet Arend-Roland**
This picture of the comet was taken on April 25, 1957, five days before Figure 7.19. The stars appear as trails, since the camera followed the moving comet. Lick Observatory photograph.

sunlight on the material in the nucleus causes its internal activity to increase. As its distance from the sun diminishes, the ices melt and the light changes from strictly a reflection of sunlight to a reflection plus fluorescence. The comet absorbs some of the solar energy and reemits it, thus producing a bright-line spectrum in addition to the dark lines of the reflected solar spectrum. Lines of the unstable radicals OH (the hydroxyl radical), CN (cyanogen), and NH and $NH_2$ (hydrides of nitrogen) have been recorded in cometary spectra. The comets also contain considerable iron. The

Figure 7.19 **The Comet Arend-Roland**
In this picture, taken on April 30, 1957, the "anti-tail," apparently ahead of the comet, is easily visible. Lick Observatory photograph.

spectacular appearance of a comet near the sun is achieved at the expense of a portion of its mass, which is being driven out from the head. Consequently, most comets that return after reasonably short intervals become rather unspectacular. A comet far from the sun probably looks more like a dirty iceberg than anything else. As it comes closer to the sun, the ices melt and the smaller particles are driven into the tail. The internal activity of the comet itself ejects the small particles from the nucleus. After ejection,

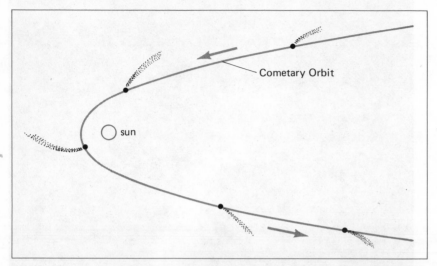

Figure 7.20 **The Tail of a Comet at Various Points in Its Orbit**
Since the tail of a comet is produced by repulsive forces (radiation from the sun and the solar wind), it points away from the sun. The orbital motion of the comet itself produces the curvature in the tail.

they are subjected to the radiation pressure of sunlight and the solar wind. The direction of both these forces is away from the sun. Therefore, comet tails generally point away from the sun regardless of the direction of motion of the comet's head (Figure 7.18). When the comet is moving away from the sun, the tail tends to be in front of it (Figure 7.19). The motion of the comet and the gravitational attraction of the sun, as well as a possible interplanetary drag by material permanently in the solar system, cause the tail to curve. The tail is not a permanent feature, but consists of moving particles traveling in orbits of their own, which may carry them completely out of the solar system (Figure 7.20).

Comets are probably members of the solar system. If they were not, they would show a preference for motion away from the general direction of the constellation Hercules, toward which the solar system is moving with a velocity of about 12 miles a second. Since they show no preference for this particular direction in space, we may assume that, whatever their source, it is attached to or moving along with the solar system. The Dutch astronomer J. H. Oort has suggested the existence of a reservoir of 100 billion or so comets at a distance of 100,000 or more AU's. When comets come near the sun, they shine with great brilliance; some are perturbed into smaller orbits; others, like the great comet of 1882 (which was easily visible in the daytime), return to great distances from the solar system.

## 7.7 THE ORIGIN OF THE SOLAR SYSTEM

Having discussed the qualitative and quantitative properties of the members of the solar system and their behavior, we now come, as a final exercise, to speculation about the origin of the solar system. It should be noted that we are not speaking of a creation, but rather of a development from a simpler state. Any theories regarding the evolution of the solar system must start somewhere, with a presumably simpler situation than that which exists at present. Moreover, they must account for all the known properties of the system.

Obviously, there are no eyewitness accounts of the beginning of the solar system. To complicate the problem further, we have no systems other than our own to examine. At present, the theories of the evolution of the stars are more solidly based than those of the origin of the solar system, because there are billions of stars available for inspection. Also, we have only our observations of the present state of the solar system to work with, whereas we have stars of all ages to observe. It is as if a panorama of stellar evolution were laid out before us in the sky. All theories, however, must be updated as objections are raised and information increases.

Let us first review the properties, particularly the regularities, of the solar system and its members. Centrally located and dominating the entire system is the sun. Because of its relatively small radius (compared with the total radius of the solar system), it possesses something less than 2 percent of the angular momentum of the system (which is equal to the sum of all the particles in the system, multiplied by their velocity and their radius of rotation from the center) although its mass is very large. The planets, because of the size of their orbits, possess about 98 percent of the system's angular momentum.

The regularities of the system are those we have studied. The planets have nearly circular orbits, small inclinations to the plane of the ecliptic, and, viewed from the north, travel from west to east around the sun. All except Venus and Uranus also rotate from west to east. There have probably been considerable evolutionary changes in the rotation of the planets over the 4 or 5 billion years that we presume the system to have existed. There is a great deal of interplanetary material in the solar system, ranging from the larger asteroids down to particles of atomic size. This interplanetary medium must be considered in any theory of evolution.

Related to the problem of the origin of the solar system are the theories of stellar evolution. Present theories assume that the stars began as clouds of gas and dust which gradually contracted. Perhaps as much as 50 percent of the material in the neighborhood of the sun is in the form of dust and gas, so plenty of material for stellar formation is still available. In any contracting cloud, a mild turbulence would be set up by the contraction; and in all probability, irregularities in gravitational distribution would form points about which accretion would take place. Turbulence and lack of symmetry could conceivably result in rotation. In any isolated system, the angular momentum can be transferred from one particle to another, much as energy is transmitted from the earth to the moon by the tides, but it cannot be created or destroyed. These are thoughts we need to bear in mind as we examine the theories held by various philosophers and scientists about the origin and development of the solar system.

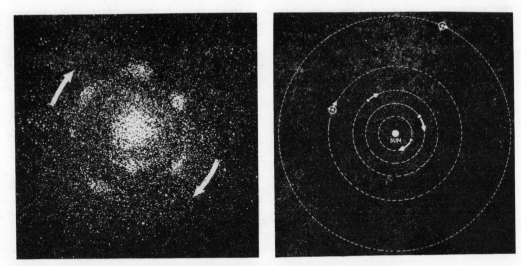

Figure 7.21 **Kant's Theory of Evolution (1755)**
The picture on the left shows a mass of gas and dust in rotation, as postulated by
Kant. The picture on the right shows what was presumed to be a later stage in
the evolution of the solar system, as clots grew by accretion to form planets and
satellites and the remainder of the nebula contracted to form the sun. Yerkes
Observatory, University of Chicago.

The earliest scientific theory regarding the evolution of our
system is what might be called the Kant-Laplace nebular hypothesis
(Figures 7.21 and 7.22). First proposed in 1755 by the philosopher
Immanuel Kant, this theory attracted no attention until the end of
the century, when Pierre Laplace, a French mathematician, formally
presented it to the scientific community. Basically, the theory was
as follows: Sometime in the remote past, a large ball of gas, many
times the size of the present solar system, was slowly rotating
about a central axis. Its temperature was low, so that the gravitational attraction of the mass prevented a significant escape of the
lighter gases at its periphery. Gradually, its own gravitational attraction caused it to contract. Because of the need to preserve
angular momentum and the decreasing average radius due to the
contraction, the velocity of its rotation increased. Centrifugal force
near the equator of the rotating mass then caused a bulge in the

**213**

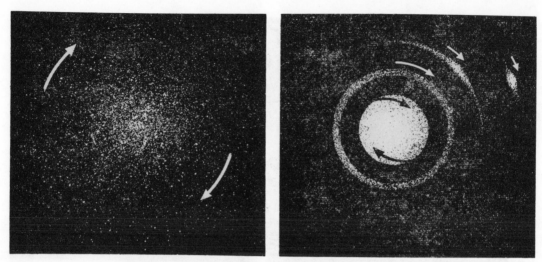

Figure 7.22 **Laplace's Theory of Evolution (1796)**
Like Kant, Laplace postulated a rotating nebula of hot gas (left). He believed,
however, that the planets were formed by rings of gas left behind as the cooling
nebula (right) shrank. The remainder of the nebula formed the sun. Yerkes Observatory,
University of Chicago.

ball of gas. When the centrifugal force had increased sufficiently,
the ring of gas that formed the bulge was left behind by the contracting mass. This ring then grew into a planet, by some process
or agency not clearly described. The central mass continued to rotate and contract, and successive rings corresponding to the successive planets were left behind.

This hypothesis survived through the nineteenth century, largely
because of the prestige of Laplace as a mathematician and scientist,
but ran into considerable difficulties, for the following reasons:
First, the information available at the beginning of the nineteenth
century about the nature of the properties of the solar system and
its members was not nearly as complete as that available at the
end of the century. Second, it had been shown that the rings necessary for the formation of the planets would not have formed of
themselves. The velocity necessary to detach such a ring from a
ball of gas was far too great. Third, even if the rings could have

214

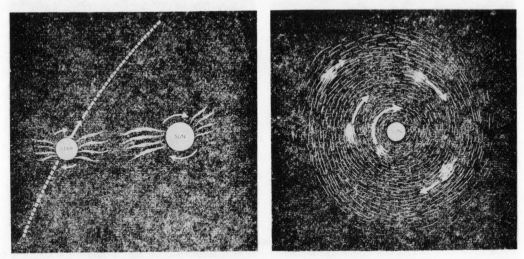

**Figure 7.23 The Chamberlain-Moulton Hypothesis (1900)**
According to this theory, the impetus for the formation of the solar system came when a passing star narrowly missed the sun, causing eruptions on both bodies, as in the picture on the left. The material ejected from the sun then condensed and coagulated to form the planets, through the process shown on the right. Planets would also form about the other star. Yerkes Observatory, University of Chicago.

formed, they would not have coalesced into planets, but would have dissipated into space. Finally, most of the angular momentum of such a formation would have remained with the central mass, and the distribution of angular momentum within the system would not approximate what we observe, namely that the planets have 98 percent of the total mass and the sun only about 2 percent. However, the Kant-Laplace theory served an admirable purpose in inspiring and stimulating scientific thought, and even its abandonment brought about further investigations into the problem.

At about the beginning of the twentieth century, another pair of closely related theories appeared. Chamberlain and Moulton in the United States hypothesized that the solar system was created by the close approach of a star to the sun (Figure 7.23). Indeed, they suggested, two solar systems could very reasonably have been created by such a phenomenon, one for each star. As the star

**215**

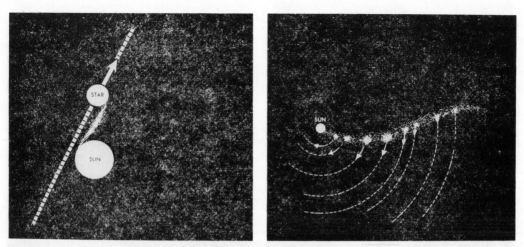

Figure 7.24  **The Jeans-Jeffries Hypothesis (1917)**
In this hypothesis, a passing star (left) sideswipes the sun, tearing out a long filament of gas. The gas cools and condenses into planets, the largest in the middle and smaller ones at either end, as in the picture on the right. Yerkes Observatory, University of Chicago.

approached the sun in a hyperbolic orbit, tidal bulges of enormous size would have been produced on both bodies. Material would have been pulled from the sun and presumably from the other star as well. Most of the material fell back into the two stars, but some coalesced into tiny particles called *planetesimals*. These particles then formed the planets by mutual gravitation and accretion. The orbit of the other star, and consequently the orbits of the planets, lay in the plane of the present ecliptic or very close to it. Circular, or nearly circular, orbits would result from the averaging out of the individual velocities of the planetesimals as they coalesced to form the larger planets. Revolution and rotation in the same direction could be accounted for by the direction of motion of the other star with respect to the sun. The rotation of the sun on an axis nearly perpendicular to the ecliptic was more difficult to explain.

At about the same time, Jeans and Jeffries in England put forth a similar theory, also based on the close approach of a star. They

216

hypothesized the expulsion from the sun of a long tidal filament tapering at the ends (Figure 7.24) and accounted in this way for the large planets' central location. Later, Henry Norris Russell examined the proposal that some of the difficulties of the other two theories could be eliminated if a grazing collision, rather than a simple close approach, were postulated. He also discussed the theory that the sun was a double star whose companion had been destroyed to form the planets.

The change from a nebular theory of the formation of the solar system to various collision theories meant a considerable revision in scientists' thinking about the universe as a whole. If the nebular theory was correct, then planets could occur as part of the normal evolution of almost any star, and solar systems should be fairly common. Close approaches of one star to another and grazing collisions, on the other hand, have a very low probability of oc- currence. The average separation of stars in the sun's neighborhood approximates that of ping-pong or tennis balls 400 to 800 miles apart. The chance of a collision between two such objects is almost zero. If the collision theory is accepted, then we must assume that solar systems are very rare indeed. However, within a few decades after the Chamberlain-Moulton and Jeans-Jeffries theories were proposed, too many impossibilities in them had become evident, and the nebular hypothesis returned to prominence, although in a considerably revised form.

Most astronomers today seem to favor the idea of a contracting ball of gas, not one that leaves rings behind to be mysteriously transformed into planets, but a turbulent aggregation with localized regions growing by accretion under their own gravitational attrac- tion. All forms of this theory seem to be very wasteful of material. It is estimated, in one case, that the original globe of material from which the earth was formed had a mass somewhere between 100 and 1000 times the mass of the final product—in other words, a mass equal to three or four times the mass of the planet Jupiter— but 99.9 percent of this material was swept away. This is reason-

Figure 7.25 **The Weizsäcker Hypothesis (1945)**
According to this theory, vortices were formed in the equatorial plane of the nebula
rotating about the sun. Accretion took place along the heavy concentric circles,
forming planets and satellites with direct rotation and revolution. Yerkes Observatory,
University of Chicago.

able, since the primeval gas would be largely hydrogen and helium,
with a trace of heavier elements. Jupiter may be large enough to
have retained all or most of the mass of the lighter two elements.
The earth, on the other hand, would have been formed by the
residual heavier elements, the hydrogen and helium having for the
most part escaped.

In 1945, Weizsäcker proposed that the planets had been formed
from the vortices of a turbulent mass of gas rotating about the sun,

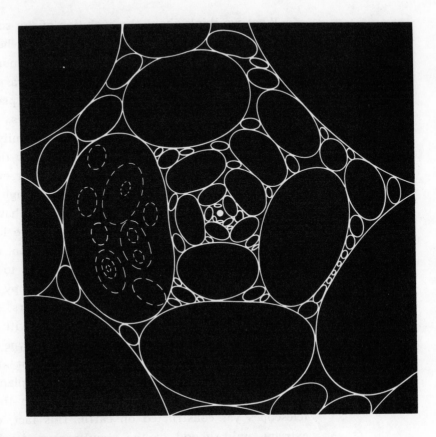

**Figure 7.26  The Kuiper Hypothesis (1950)**
Like Weizsäcker, Kuiper suggested a pattern of turbulence in the nebula about the sun. However, he theorized that the planets and satellites formed from "photoplanets." Yerkes Observatory, University of Chicago.

as in Figure 7.25. By a judicious choice of the turbulent eddies, he could approximate the positions for the planets predicted by Bode's law. The theory was not universally accepted, nor was a similar theory proposed by Kuiper (Figure 7.26). A great deal of information has yet to be gathered before we can arrive at a definitive theory of the evolution of solar systems. It is hoped that future space vehicles and astronauts will help to provide this information.

**219**

What we think about the evolution of the solar system affects what we think about the possibility of life elsewhere throughout the universe. If the formation of planets is a normal sequence of events in the evolution of a star, or even of a particular class of star, there should be, as we have noted, many planetary systems in our galaxy and many more in other galaxies. However, it does not follow that every star has a planetary family, nor would such a planetary family necessarily be hospitable to the development of life. As we shall see when we consider stellar evolution, the massive, hot stars simply do not live long enough for the complex molecules necessary to the chemistry of life to have much chance of developing. Cool stars with a small mass have long lives, but the *range of habitability* about them (the region within which the orbit of a planet would have to be for life to develop) is very narrow. Stars similar to the sun strike a happy medium between these two extremes of very hot and cool stars, so that we might have better luck in finding a solar system with life forms if we were able to examine stars of the general type and temperature of the sun more closely. However, so distant are the stars, and so small is the largest hypothetically possible planet, that no planet of even the closest star could be seen with the most powerful telescopes yet constructed on earth. This fact, coupled with the extreme unlikelihood of the simultaneous development of communication ability, makes the hope of directly detecting the existence of life on other planets extremely remote.

# QUESTIONS

(1) When was the first asteroid discovered?     (Section 7.1)

(2) Describe a method of discovering asteroids using a telescope and photography.                                    (Section 7.1)

(3) What are the Trojan asteroids?            (Section 7.2)

(4) How are Cassini's divisions similar to the Kirkwood gaps?
                                               (Section 7.2)

(5) Briefly describe one possible theory of the origin of asteroids.                                    (Section 7.3)

(6) What is the difference between a meteor, a meteoroid, and a meteorite?                              (Section 7.4)

(7) Why do we see more meteors after midnight than before?
                                               (Section 7.4)

(8) What is the zodiacal light?            (Section 7.4)

(9) Describe the three categories of meteorites.
                                               (Section 7.5)

(10) What is a tektite?                    (Section 7.5)

(11) Describe the changes in the appearance of a comet as it orbits the sun.                                  (Section 7.6)

(12) Why are comets thought to be members of the solar system?                                     (Section 7.6)

(13) Describe Weizsäcker's theory of the origin of the solar system.                                     (Section 7.7)

# CHAPTER 8
# THE SUN

The sun is the dominant member of the solar system. Its importance to the earth is inestimable. It is the major source of heat and light for the human race. It is by far the closest star to the earth, and it is the only star that, to the naked eye, does not look like a point of light. Thus, the study of the sun constitutes a bridge to the wide field of stellar astronomy.

# 8.1 THE PHYSICAL PROPERTIES AND DIMENSIONS OF THE SUN

The sun (Figure 8.1) is a ball of gas at an intensely high temperature, with a diameter of 864,000 miles, 109 times that of the earth. (The solar diameter is determined from our knowledge that the sun is about 93 million miles away and subtends an angle of about $\frac{1}{2}°$.) If the earth were represented by a ball about the size of a dime, the sun would be 6 feet in diameter and the distance between the two bodies would be approximately 660 feet. Figure 8.2 shows the size of the sun compared to the moon's orbit. The volume of the sun is $1\frac{1}{3}$ million times the volume of the earth, but its mass is only about 300,000 times that of the earth, or $2 \times 10^{33}$ grams. These figures result in a density about a quarter that of the earth, or 1.4 times the density of water, which is rather close to the average density of Jupiter.

Since the sun is gaseous, it is highly compressed toward the center. Thus the densities of the parts of the sun we can see are considerably lower than the average density. The sun becomes opaque as we look into its atmosphere, but even at the visible surface the density is extremely low.

The size of the sun and its mass can be used to compute its velocity of escape and surface gravity. The velocity of escape is 56 times that of the earth, while the surface gravity is approximately 28 times that of the earth. A 100-pound object on the surface of the earth would weigh 2800 pounds on the surface of the sun. Because of its extremely high temperatures, the velocities of the particles of the sun's atmosphere are such that escape is not particularly difficult. Nonetheless, the sun has retained and replenished its atmosphere throughout its lifetime.

A *gauss* (G) is a unit of measure of a magnetic field. The sun has a general magnetic field of 1 G, slightly stronger than the magnetic field of the earth, which is about 0.7 G. *Magnetograms*, graphic representations of magnetic scans of the solar magnetic field, that confirm this particular feature of the sun have been produced by H. D. and H. W. Babcock. However, there is still a great deal to

223

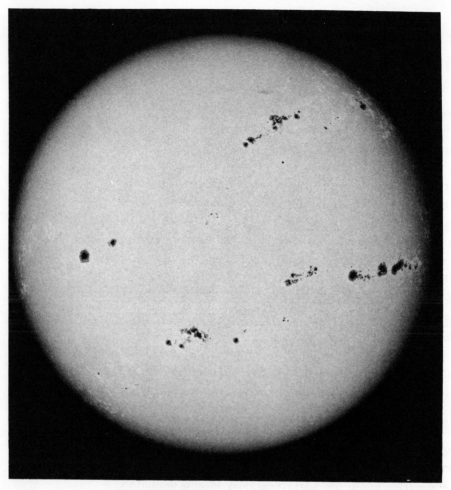

Figure 8.1 **The Sun**
This photograph was taken near the sunspot maximum, on December 21, 1957.
Photograph from the Hale Observatories.

be learned about the sun's magnetic field, and research in this area is continuing.

Although solar observations may be carried out with ordinary reflecting or refracting telescopes by means of special eyepieces, projection screens, or the attachment of other equipment, the sun can be observed more efficiently with telescopes designed particularly for this purpose. Because the sun delivers billions of times more light to the earth's surface than the brightest of the other stars, solar telescopes can afford to waste light. They are also de-

224

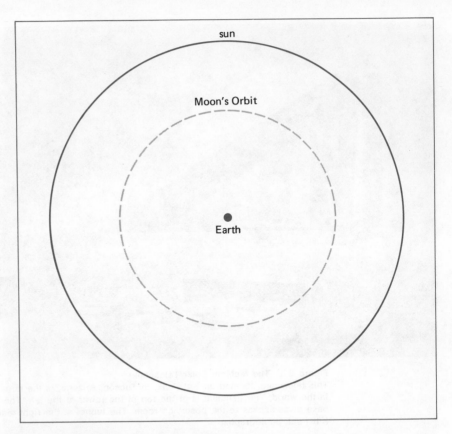

Figure 8.2 **The Size of the Sun Compared to the Moon's Orbit**
If the earth were placed at the center of a hollow sphere the size of the sun and
the moon retained its present orbit, the moon would be a little over halfway between
the earth and the surface of the sphere. The sun itself is about 109 times the
diameter of the earth.

signed to dissipate the heat of the solar radiation to ensure a steady,
clear image. Among the most noteworthy of the special solar telescopes are the 60- and 150-foot towers and the horizontal telescope
(though this is rarely used) at Mt. Wilson. These have tilted,
rotating mirrors called *coelostats* in front of their objective lenses
that cause the sun to appear to "stand still." The objective plane
mirror rotates at 7.5° per hour and reflects the light of the
moving sun onto a stationary lens. This lens forms the solar
image at its focal point. In the towers, the beam is vertical and
the image is formed at ground level. Because of the abundance of
light, the extra mirror can be used without a significant loss of

Figure 8.3 **The McMath Solar Telescope**
This telescope, located on Kitt Peak, in Tucson, Arizona, is the largest of its kind in the world. The heliostat is at the top of the tower at the left. The wide tunnel at the base gives access to the observing room. The tunnel at the right leads to the shops. Kitt Peak photograph.

light. The advantage of the coelostat is that everything except the moving mirror is fixed. No long, heavy telescope need be moved to follow the sun. The solar image may be studied at the focal point, or the light may be passed through a slit into a well and reflected back to the working level from a grating to form the solar spectrum. At the Kitt Peak Observatory (Figure 8.3), the main shaft of the solar telescope above the ground is parallel to the axis of the earth. This makes the operation of the coelostat somewhat more convenient.

## 8.2 THE PARTS OF THE SUN

The names of the parts of the sun are somewhat arbitrary, and the boundaries between them are indefinite. Everything below the

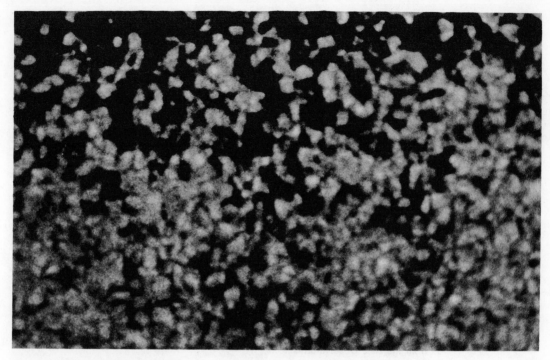

Figure 8.4 **The Surface of the Sun**
This high-resolution photograph shows the granular structure of the solar surface. The granules are about 300 miles long. Photograph from the Hale Observatories.

deepest distance we can penetrate optically is called the *interior* of the sun. Its nature must be inferred from the application of the laws of physics and our observations of the exterior. Since the sun is a typical star, we will leave the discussion of its interior to Chapter 9, where stellar interiors in general are studied.

The deepest level of the sun to which the eye can penetrate, and the lower limit of transparency, is called the *photosphere* (sphere of light). The photosphere is the sun's visible surface. It is not, however, a discrete surface like that on the earth, the moon, or Mars. The photosphere is probably about 250 miles thick. It is deepest in the center of the visible solar disk because the center is more transparent. Near the edge, or limb, of the sun, we are unable to see as far into the gases near the sun's surface; consequently the apparent level of the photosphere is less deep.

The photosphere of the sun is not featureless; it exhibits considerable granulation, as shown in Figure 8.4. The granules in

227

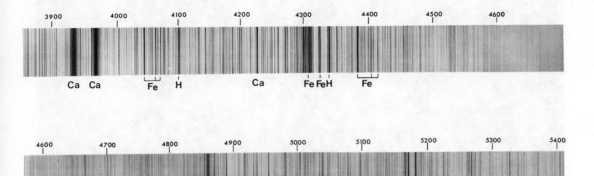

**Figure 8.5  A Portion of the Solar Spectrum**
This photograph shows a portion of the visible region of the solar spectrum from about 3900 Å to 5400 Å. The numbers on the top of the spectrum give the wavelengths; the letters underneath indicate the chemical elements producing the lines. Photograph from the Hale Observatories.

the sun's surface are about 300 miles across and are near the limit of visibility with earthly telescopes. The best pictures of them were taken using telescopes carried by balloons to elevations of about 80,000 feet above the earth's surface. These photographs show the granulations to be updrafts of hot material from the interior of the sun. The darker interstices between the granules are downdrafts.

The opacity of the photosphere is produced by negative hydrogen atoms (atoms that have acquired an extra electron) in the plasma that are opaque to the transmission of light. Immediately above the opaque layer of the photosphere is the *reversing layer*. This is a thin layer transparent to most, but not all, solar radiation. The dark lines of the solar spectrum (Figure 8.5) are produced in this layer. At this level, atoms and a few molecules extract discrete wavelengths from the continuous background of photosphere below, causing dark lines in the spectrum. It is from these lines that we have gathered a great deal of information about the sun.

Above the reversing layer is the *chromosphere* (sphere of

color), about 5000 miles thick. Above the chromosphere, and extending well above the surface of the sun, are the *prominences*, great protuberances of gas that surge tens or even hundreds of thousands of miles above the sun's surface. Finally, extending well beyond the surface of the sun, we find the *corona* (crown). As we have pointed out, however, the sun is a gas, and boundaries of gas clouds are not discrete. The boundaries between the reversing layer and the chromosphere and between the chromosphere and the corona, and even the outer boundary of the corona, are not sharply defined.

## 8.3 THE ENERGY OF THE SUN

Even the most casual observer realizes that for the sun to produce the heat and light it does at a distance of 93 million miles, its total energy output (or *luminosity*) must be very large. The sun shines at the expense of its own mass, by converting hydrogen atoms into atoms of helium. The helium atoms have a slightly lower mass than the hydrogen atoms from which they were formed, and the excess mass is converted into energy in accordance with the Einsteinian formula

$$E = mc^2$$

where $E$ is energy, $m$ is mass, and $c$ is the velocity of light. Six hundred million tons of hydrogen are converted into helium each second to supply the sun with energy. Although this is an enormous amount of energy by terrestrial standards, the sun has enough mass in the form of hydrogen atoms to form helium for a period of about 10 billion years.

The quantity of energy released by the sun is determined by measuring the *solar constant*, the amount of energy received at the outer atmosphere of the earth. By measuring this quantity, and knowing the distance to the sun, we can calculate the total energy it produces, which turns out to be about $4 \times 10^{23}$ kilowatts. So far as we know, radiation has been emitted from the sun at essentially this rate for approximately 5 billion years and will continue to be emitted for another 5 billion years. The source of this energy was a mystery for many years, until the development of thermonuclear physics made a reasonable explanation possible. Various theories had been considered before, of course. One, proposed by Hermann von Helmholtz, suggested that the sun was contracting. A contracting gas is heated, and it is easy to show that the necessary rate of contraction of the solar surface is not unthinkable. However, the sun could not have lived to its present age contracting at this rate. Another theory suggested that the accretion of meteors on the sun's surface enabled it to radiate at a constant rate.

## 8.4 THE TEMPERATURE OF THE SUN

We can calculate the temperature of the photosphere of the sun by applying various laws of thermodynamics. The most general of these is *Planck's law*, which tells us the relative strength, or amount, of radiation at each wavelength for a particular absolute temperature (Figure 8.6). Using this law, we can graph the quantity and distribution of energy emitted by a *blackbody* at that temperature. (A blackbody, or perfect radiator, is a theoretical body that absorbs and reemits, completely, all electromagnetic radiation falling on it. No such body exists in nature, but the sun and other stars approximate this condition closely enough so that for certain purposes they may be assumed to be blackbodies.) Graphing emissions at various temperatures, as in Figure 8.7, shows that any blackbody radiates some energy at all wavelengths, that a hotter blackbody emits more energy at each wavelength than a cooler one, and that the wavelength at which the most energy is emitted grows shorter (bluer) as the temperature of the blackbody rises. This latter phenomenon is easily illustrated by heating a bar of iron.

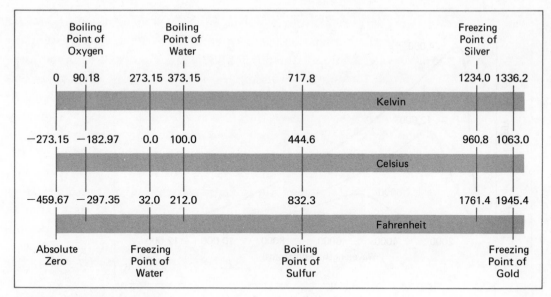

**Figure 8.6 Temperature Scales**
This diagram shows the three most commonly used temperature scales. Scientists prefer the Kelvin scale, which begins at absolute zero. It is impossible for anything to be colder than 0°K.

At first, the bar glows a dull red. As its temperature increases, it glows more brightly and becomes yellow or orange. If the bar could be prevented from melting at very high temperatures, it would glow a brilliant blue. The wavelength of maximum emission (the peak of a curve in Figure 8.7) is called $\lambda_{max}$.

By plotting theoretical Planck curves for selected temperatures and matching observed solar or stellar radiation curves to them, temperatures of the sun and stars can be obtained. Of course, an adjustment must be made for absorption and for the fact that the observed objects are not perfect blackbodies.

The shift of $\lambda_{max}$ toward the blue end of the spectrum and the increase in energy output as the temperature increases can be derived from Planck's law. For example, *Wien's law* states that $\lambda_{max}$ is inversely proportional to the absolute temperature of the blackbody. This law has the following mathematical form

**231**

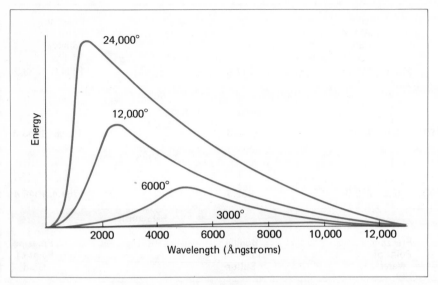

**Figure 8.7  Blackbody Radiation Curves**
This graph shows the Planck curves, which tell us how much energy is radiated at various wavelengths by an ideal blackbody. As the body becomes hotter, more radiation is emitted, in accordance with Stephan's law. In addition, as the temperature of the body rises, the wavelength at which most of the radiation is emitted moves further and further toward the blue, in accordance with Wien's law.

$$\lambda_{max} = \frac{0.2897}{T}$$

where $\lambda_{max}$ is in centimeters and $T$ is in degrees Kelvin.

The *Stefan-Boltzmann law* says that the total energy, E, produced per second per square centimeter is proportional to the fourth power of the absolute temperature $T$; that is,

$$E = aT^4$$

where $a$ is a constant with a specific value. In other words, if the temperature of a radiating body doubles, it will emit 16 times as much energy.

The temperature of the surface of the sun is 6000°K according to Planck's law, 6080°K according to Wien's law, and 5980°K according to the Stefan-Boltzmann law. The slight differences in the calculated temperatures occur because the sun is not a perfect radiator and because of the dark lines in the solar spectrum. Basically, the temperature of the photosphere appears to be about 6000°K. Sunspots, however, have an observed temperature of about 4600°K. Moreover, for the sun to produce the amount of observed radiation it does, the temperature at the center must be approximately 13 to 15 million degrees.

The radiation we have been discussing thus far is that in the vicinity of the visible spectrum. The sun is also a source of radio energy in wavelengths of about 1 to 15 centimeters. Various wavelengths are produced at different levels in the sun's atmosphere. Radiation with a wavelength in the 1-centimeter range comes from the lower chromosphere; radiation with a wavelength of 15 centimeters comes from the upper corona. Hence, by tuning to different wavelengths, we can obtain a radio picture of various levels of the sun's atmosphere.

The fluctuation in radiant energy from the sun in the visual and infrared range is relatively small. In the radio range, however, the variability is greater. When the surface of the sun is relatively inactive, we observe what is called the *quiet sun*. The amount of radiation received in the radio range at this time is about what we would expect from a body with the sun's temperature. However, during times of great solar activity, when sunspots are numerous and flares are frequent, large outbursts of radiant energy in the radio range are also observed. The sun during such periods is called the *active sun*.

## 8.5 SUNSPOTS

Sunspots were first observed telescopically by Galileo, but long before the invention of the telescope, spots large enough to be observed with the naked eye were occasionally recorded by ancient astronomers. A telescopic examination of the solar disk will almost

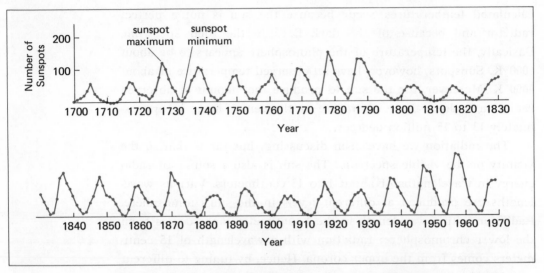

**Figure 8.8** **The Variation in the Number of Sunspots During an 11-Year Cycle**
This graph shows the average daily number of spots counted on the surface of the sun since 1700. Note the steady rise and then fall of the number of spots in each cycle. The approximate period of variation is 11 years.

always reveal at least a few spots. In 1851, Heinrich Schwabe found that the number of spots on the sun fluctuated in cycles of approximately 11 years. As Figure 8.8 shows, observations to date support this contention. The period during each cycle when the number of spots is greatest is called the *sunspot maximum*. A period when there are few or no spots on the sun is called a *sunspot minimum*.

To the eye, sunspots appear simply as darker areas on the solar surface. Telescopic and photographic observations show, however, that typical sunspots have two parts (Figures 8.9 and 8.10). The inner, or central, region is the much darker of the two and is called the *umbra*. The lighter outer region surrounding the umbra is called the *penumbra*. (The student will notice the identity of these names with those given to the parts of the moon's and earth's shadows discussed in Section 5.13.) Sunspots range in size from less than 500 miles to almost 100,000 miles in diameter. They usually appear

234

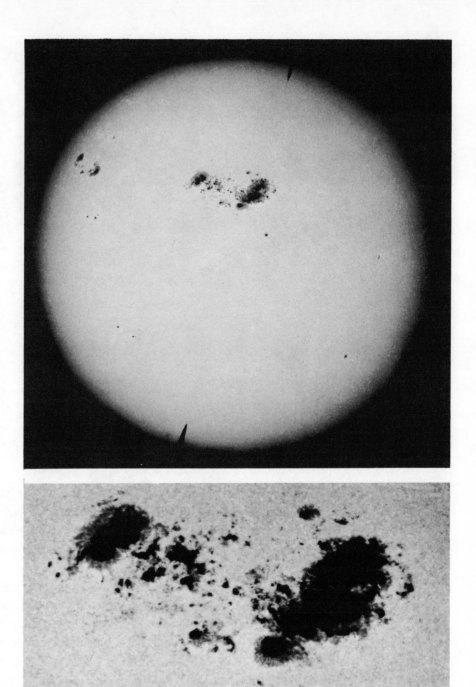

Figure 8.9  **The Sun and a Large Sunspot Group**
This photograph, taken on April 7, 1947, shows the whole solar disk. Below is an
enlarged view of a large spot group. Photographs from the Hale Observatories.

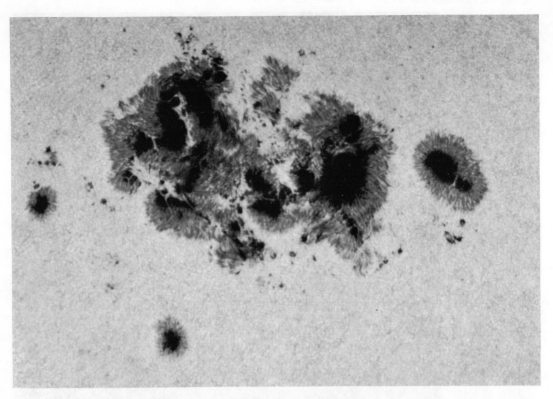

Figure 8.10 **A Large Sunspot Group**
This picture of a large sunspot group was taken on May 17, 1951. The leader is at the right. Photograph from the Hale Observatories.

in groups, with a principal spot called the *leader* preceding the group and a *follower* spot bringing up the rear. A large group will have many smaller spots between these two. A group may be spread over a large region, but no group ever covers more than a small fraction of the total surface of the sun.

At a sunspot minimum, a new cycle of spots is about to begin. They appear first on both sides of the sun's equator in a solar latitude of about 30° to 35°. Spots are rarely seen farther from the equator than this latitude. The latitude of the spots does not change, but the zones in which they are most numerous gradually drift toward the equator during the sunspot cycle. When the cycle ends, the last spots visible are about 5° to 8° on either side of the equator. At about the time that the last spots of one cycle end, the first of a new cycle of spots appears in high latitudes, and the whole process

236

is repeated. Recent sunspot minimums occurred in 1933, 1944, 1954, and 1966.

An area of the sun appears as dark because of a difference in temperature. As we have noted, the temperature of the photosphere is about 6000°K and that of the umbra of a sunspot is about 4600°K. If we could lift the sunspot away from the surface of the sun and view it against a dark background, it would be redder than the sun, because of its lower temperature, but not as dark as it appears in contrast to the photosphere. An examination of sunspots with a spectroscope reveals a different line intensity than that of the direct solar spectrum.

Sunspots were formerly regarded as highly turbulent regions, or "storms on the sun's surface." It now appears that sunspots are not turbulent, but calm. They have strong magnetic fields, which would tend to make them rigid and less fluid than the surface of the sun itself. The average magnetic field of the sun does not exceed a few gauss. Fields from 100 to 4000 G have been measured in sunspots. A strong magnetic field applied to a beam of light splits the dark lines in its spectrum. This is known as the Zeeman effect (Section 3.8), after its discoverer. An examination of the spectrum of sunspots reveals just such split lines.

The polarity of any spot depends on the direction of revolution of the gases in it. The leader and follower spots in any group are of opposite polarity. The leader spot in a group in the Southern Hemisphere is of opposite polarity from the leader spot in a group of the Northern Hemisphere in the same cycle. Finally, the polarities of all the spots are reversed when the cycle ends and a new cycle begins. In a sense, then, we should perhaps think of the sunspot period as 22 years instead of 11 years.

Solar activity correlated with the sunspot cycle appears to have some connection with *geomagnetic storms* on the earth. That is, during the time of maximum sunspot activity, magnetic storms, fluctuations in the ionosphere, fade-outs of short-wave transmissions, and disruptions of teletype and telephone service are at a maximum also.

Another phenomenon that seems closely connected with the number of sunspots is the aurora (the northern and southern lights). Auroras are most numerous during a sunspot maximum and are frequently seen as far south as the southern United States during this period. During a sunspot minimum, they are rarely seen south of the Canadian border.

Attempts to correlate other earthly phenomena with the sunspot cycle have not been too successful. However, scientists in Arizona have for some years investigated a correlation between sunspot cycles and variations in the spacing of tree rings. Whether this spacing is an indication of variability in rainfall, temperature, evaporation, or some other factor is not certain. It may well be that there is a correlation between the sunspot cycle and such terrestrial phenomena.

## 8.6 THE ROTATION OF THE SUN

A typical sunspot appears at the eastern limb of the sun and is first noticed almost on edge. Each day it appears to move farther west, until after about a week it is in the center of the solar disk. It continues its apparent trek westward and about a week later disappears around the western limb. If it lasts long enough, in two more weeks it will appear again at the eastern limb. Actually, of course, it is the sun, and not the spot, that is moving, rotating on its axis once in about a month. Unlike the solid earth, the sun does not rotate on its axis in the same period in all latitudes. An examination of sunspots near the solar equator shows the period of rotation of the sun there to be about 25 days. As one moves farther north or south, the period of rotation increases.

Another method of determining the rotation period of the sun is to allow the solar image to fall on the slit of a spectroscope and measure the Doppler shift of the dark lines. If the slit is placed on the eastern, or approaching, limb of the sun, the lines will be shifted to the violet end of the spectrum; if it is placed on the western limb, the shift will be toward the red. The extent of the shifts gives us the sun's rotational velocity. Knowing the circumference of the sun in the latitude observed, we can then compute the time necessary for one complete rotation. The results of such calculations are consistent with the data obtained from observations of sunspots and enable us to determine rotational periods in latitudes above 35°, where spots do not appear. The period near the poles is 35 days, and the axis of rotation is inclined 7°10' to the ecliptic.

## 8.7 WHAT THE SPECTROHELIOGRAPH REVEALS

In the early 1890s, George Hale in the U.S. and Henri Alexandre Deslandres in Europe independently developed the *spectroheliograph*, an extension of the spectrograph. Light from the sun is first passed through the spectrograph to form a spectrum. This is done by permitting the telescopic image of the sun to fall on the slit of the spectrograph. A spectrum is formed, but instead of a photograph being taken, another slit is placed in front of the spectral light. This slit is adjusted so that a particularly prominent dark line, such as a hydrogen or calcium line, falls on the second slit. The two slits are then moved at the same rate so that the first scans the solar image and the second moves with respect to the photographic film or plate behind it but keeps in perfect step with the moving image of the dark line. In effect, then, the picture produced on the film or plate is a scanned image of the solar disk taken in the light of the dark line. (Actually, there is a considerable amount of light coming through the regions of the spectrum where these "dark" lines appear. They appear dark only in contrast with the bright continuous background.) The resulting photograph is a picture of the sun at a particular level and in the light of a particular element. Obviously, different types of pictures are obtained, depending on which line is used. Even when the same line is used, different

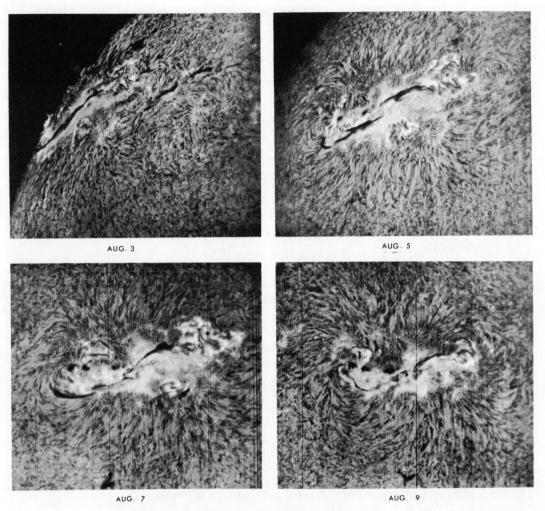

AUG. 3

AUG. 5

AUG. 7

AUG. 9

Figure 8.11 **A Series of Spectroheliograms**
These pictures of a section of the sun were taken in red hydrogen light from
August 3 to August 9, 1915. Photographs from the Hale Observatories.

results are obtained depending on whether the second slit in the spectroheliograph is placed on the center of the line or on the edge, known respectively as the *core* and the *wing*. Low-level spectroheliograms resemble direct photographs of the sun, whereas high-level spectroheliograms are quite different. It was from such pictures that the vortex structure above sunspots was found to be cyclonic, although the structure of the spot itself is probably controlled magnetically.

240

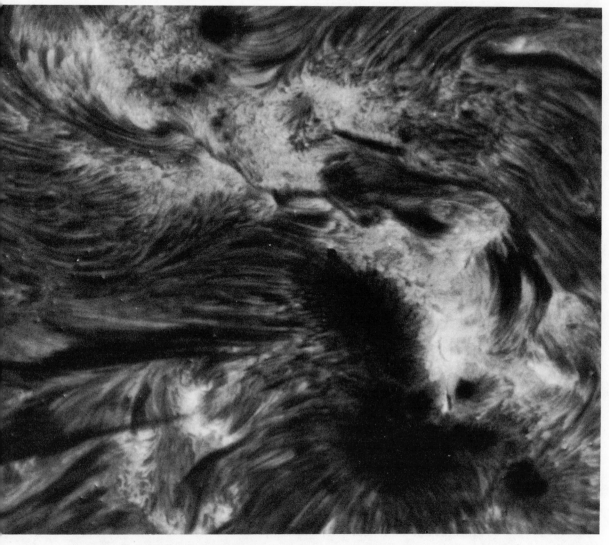

Figure 8.12 **An Active Area of the Sun**
This spectroheliogram was taken in red hydrogen light on August 25, 1971.
Photograph from Big Bear Solar Observatory.

Regions of high luminosity around sunspots are quite frequently observed with spectrohelioscopes. Formerly called *flocculi*, they have more recently been named *plages*. In white light, *faculae* (little torches) are also frequently observed. Spectroheliograms taken in 1915 and 1971 are shown in Figures 8.11 and 8.12.

Instead of a photographic plate, an eyepiece can be attached to the spectroheliograph. It is then called a *spectrohelioscope*. If the two slits of a spectrohelioscope are oscillated at the same rate fast enough, so that persistence of vision gives an image to the human eye, the results may be viewed directly. Rotating prisms may also be used instead of the oscillating slits and are smoother in operation. With the spectrohelioscope, it is possible to observe both the sun in the light of a single element and the prominences.

Carrying the process one step farther, stop-action motion pictures have been made at various solar observatories using what is called a *spectroheliokinetograph*. Exciting developments on the solar surface and in the atmosphere that take hours of real time can be viewed in a few minutes with this machine.

## 8.8 THE ATMOSPHERE OF THE SUN

The spectroheliograph has enabled a detailed study to be made of the atmosphere of the sun. Everything above the photosphere is considered part of the solar atmosphere. The lowest level of the atmosphere is the reversing layer, in which the dark lines of the spectrum are produced. These dark lines identify the chemical elements in the sun's atmosphere. Lines of 67 elements have been found in the solar spectrum. The missing elements are either rare, both in the stars and on the earth, or do not produce lines at the temperature of the sun's surface. Almost all are in atomic form, although several molecules have been identified, mostly in sunspots, where the temperature is lower. The spectrum of the sun indicates that it is probably three-quarters hydrogen. Most of the remaining one-quarter is probably helium. Only a very small part of the sun's mass, at most a few percent, consists of other elements.

In 1868, astronomers watching an eclipse of the sun observed what has since been named the *flash spectrum* (Figure 8.13), which is the spectrum of the chromosphere. When the disk of the moon has covered the photosphere, there are a few seconds during which

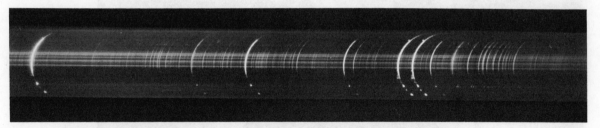

Figure 8.13 **The Flash Spectrum**
Just before or after totality, the thin crescent of the exposed chromosphere produces this spectrogram. The lengths of the individual crescents indicate the heights in the chromosphere at which the lines are produced. Photograph from the Hale Observatories.

a slender crescent of the chromosphere remains visible. No longer dominated by the more brilliant photosphere, it appears as a red flash. This can be observed with the eye, but if the light is allowed to fall on a spectroscope, a spectrum results. Since the slender crescent is already in the form of a line, no slit is necessary in a spectroscope. The resulting spectrum is that of the chromosphere. It contains bright rather than dark lines because it is an emission spectrum rather than an absorption spectrum. The gases of the chromosphere absorb the white light of the sun, which contains all colors, but reemit only the wavelengths associated with the elements (and the particular states of the elements) they contain. The resulting spectrum contains crescents rather than straight lines. The length of an individual crescent corresponds directly to the height in the chromosphere at which the element appears. A study of the chromosphere reveals that although it is 5000 miles thick, it is essentially transparent to visible light. It appears red during a solar eclipse and when observed with the spectrohelioscope because of the predominance of light produced by the hydrogen alpha line in the red region of the spectrum. The temperature of the upper chromosphere appears to be approximately 100,000°K.

Outside the chromosphere is the sun's outer atmosphere, or *corona*. The solar corona can be seen whenever there is a total solar eclipse, but at no other time without special instruments. It was ob-

Figure 8.14 **The Eclipse of June 8, 1918**
This picture shows the sunspot maximum corona. Photograph from the Hale Observatories.

served by Kepler and his predecessors and has been studied for many years. Its spectrum was first obtained in 1869. In 1930, Bernard Lyot perfected the *coronagraph*, an instrument which enables observers, particularly in high altitudes and when the atmosphere is very clear, to observe parts of the corona without the benefit of a solar eclipse. Most solar observatories today possess a coronagraph.

The corona is divided into two portions, the inner and outer

Figure 8.15 **The Eclipse of August 31, 1932**
This picture shows the sunspot minimum corona. Lick Observatory photograph.

corona. The outer regions are extremely tenuous, and even the inner corona has a very low density. Under such conditions, atoms do not react as they do under laboratory conditions, and lines are observed in the spectrum of the corona that have not been duplicated under terrestrial conditions.

The form of the corona changes with the sunspot cycle. At eclipses near a sunspot maximum (Figure 8.14), the corona has a circular symmetry like that of a sunflower. The corona at a sunspot minimum (Figure 8.15), on the other hand, has long equatorial

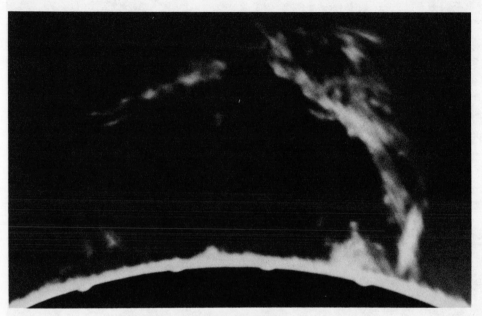

Figure 8.16 **A Solar Prominence**
This large prominence, photographed in violet calcium light on July 2, 1957, is
205,000 miles high. Photograph from the Hale Observatories.

streamers and less prominent polar tufts. The gases of the corona
seem to follow a threadlike pattern similar to the lines of force of
a magnetic field. There is no outer boundary to the corona, but during an eclipse near a sunspot minimum the long equatorial streamers may be seen to extend 2 solar diameters from the limb of the
moon. At sunspot maximum, the circularly symmetrical corona
rarely exceeds 1 solar diameter in any direction. In any event, there
is no definite boundary between the corona and empty space.

Another feature of the solar atmosphere is the *prominences*
(Figures 8.16 and 8.17). These flamelike structures that rise as high
as a million miles above the surface of the sun are classified as
quiescent or eruptive. They may be viewed around the edge of the
sun beyond the limb of the moon during solar eclipses or at any
time with a spectroheliograph. The quiescent type, as their name

246

Figure 8.17 **A Solar Prominence**
This remarkable photograph of a prominence, taken in red hydrogen light, was obtained on March 31, 1971, at the Big Bear Solar Observatory near Los Angeles. Photograph from Big Bear Solar Observatory.

implies, do not seem to move or change greatly. Their appearance is flamelike, although nothing on the sun is actually burning in the traditional sense. Eruptive prominences take on various forms, and stop-action motion pictures of them are spectacular. Some have achieved a velocity as high as 450 miles a second and have erupted above the surface of the sun in arches and flares as high as a million miles.

We must distinguish between the material of the prominence and the prominence-forming agency. When prominences are viewed in motion pictures, although the prominence itself appears to be moving outward from the sun, the material forming the prominence frequently appears to be falling toward the surface of the sun, curving as if following magnetic lines of force, and sometimes appearing to fall into the region of sunspots. Although prominences have been

247

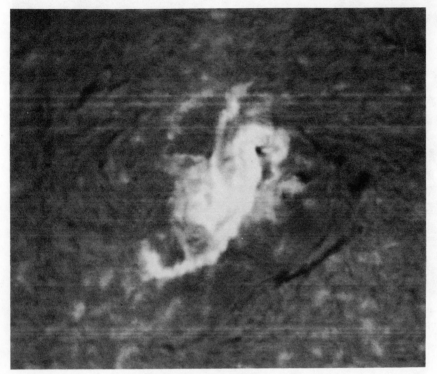

Figure 8.18 **A Solar Flare**
This flare was photographed in a hydrogen alpha line at 6556 Å on July 16, 1959.
Photograph from the Hale Observatories.

studied for many years, there is a great deal still to be learned about them.

*Solar flares* (Figure 8.18), which frequently occur near sunspots, are emissions of energy, frequently in large amounts, from the solar surface. There is a high incidence of ultraviolet solar radiation at the time of flares.

One final phenomenon associated with the sun, and worth mentioning, is the *solar wind*. The solar wind consists of charged atomic particles, streaming out from the sun in all directions. Pioneer 9 has detected it as far away from the sun as the orbit of Mars and even beyond the asteroid belt.

# QUESTIONS

(1) What is a coelostat? (Section 8.1)

(2) Briefly describe the various parts of the solar atmosphere.
(Section 8.2)

(3) How much hydrogen is converted into helium each second to provide the sun with energy? (Section 8.3)

(4) What does Planck's law tell us? (Section 8.4)

(5) What is Wien's law? (Section 8.4)

(6) What is the Stefan-Boltzmann law? (Section 8.4)

(7) What does a sunspot look like? (Section 8.5)

(8) What terrestrial phenomena are affected by sunspots?
(Section 8.5)

(9) What is the rotational period of the sun at the equator? at the poles? (Section 8.6)

(10) How does the corona change with the sunspot cycle?
(Section 8.8)

(11) What is a solar prominence? (Section 8.8)

# CHAPTER 9
# THE STARS AS INDIVIDUALS

In this chapter, we shall consider the characteristics of the stars, the building blocks of our galaxy and of the other galaxies as well. We have already studied the sun, the only star close to the earth. It has supplied a bridge to knowledge about the stars, but it is representative of only one type of star. In later chapters, we shall study the interrelationship of stars in multiple systems, clusters, and the star clouds of the Milky Way.

There are a variety of things we want to know about an individual star. Its distance from us is of paramount importance. We also want to know how it moves, how its intrinsic luminosity compares with that of the sun, the temperature of its surface, the nature and extent of its atmosphere, the constancy or variability of its radiation, and many other things.

## 9.1 STELLAR DISTANCES

As we mentioned in Chapter 4, Tycho Brahe sought to prove the truth or falsehood of Copernicus' explanation of the solar system by measuring the parallactic shift, if any, of the stars. He was unable to do so because the stars were too distant for his instruments to detect their parallaxes. It was over two and a quarter centuries after his death before a stellar parallax was first observed, by Friedrich Bessel in the 1830s. Since 1830, the investigation of the distances of celestial objects has been a major activity of many astronomers.

Bessel was able to obtain a measurement of stellar parallax by observing the same star at 6-month intervals. In 6 months, the earth moves from one side of its orbit to the other, a distance of about 186 million miles or 2 AU's. The line between these two points forms a triangle from which the star's distance can be inferred (Figure 9.1). Nearby stars obviously have larger parallaxes than more distant ones, and the first stars to be measured were those whose parallax was large. The total shift in the image of a star during a 6-month period is twice its parallax, the parallax being the maximum angle subtended by the radius of the earth's orbit as seen from the star. Two photographs taken 6 months apart are not enough (Figure 9.2). In order to eliminate all the variables, such as the star's own motion, a minimum of 18 photographs is necessary. The angle through which the closest star is shifted because of the 186-million-mile motion of the earth is surprisingly small. A right triangle 57 inches long and 1 inch wide at the base has an angle of 1° where the hypotenuse and the long side meet. One second of arc (1″), is $\frac{1}{3600}$ of this angle of 1°. The closest star in the sky, appropriately called Proxima, has a parallax of less than 1″. To have a parallax as large as 1″, a star would have to be about 200,000 AU's from the sun. Since Proxima's parallax is about 0″.78, is it even farther away.

Expressing distances to the stars required either extremely large numbers or a new unit of length. There are two units of length generally used for measuring stellar distances. One is the *light year*. Light travels about 186,300 miles per second. Since there are approximately 32 million seconds in one year, light travels 5.88 trillion

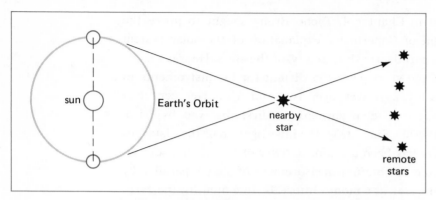

**Figure 9.1 Stellar Parallax due to the Earth's Motion**
As the earth goes around the sun, nearby stars appear to move back and forth with respect to more distant stars.

miles in a year. Another handy unit of measurement is the *parsec*, which is the distance a star would be if its parallax were 1″. This unit is equal to about three and a quarter light years, or 19.2 trillion miles. We shall use the parsec (abbreviated pc), the kiloparsec, kpc (1000 pc), and the megaparsec, mpc (1 million pc) throughout the remainder of this book.

By our definition, a star whose parallax, *p*, is 1″ has a distance, *d*, of 1 pc. Doubling the distance changes the parallax to half its former value, that is, to 0 ″5. A star 10 pc from the observer has a parallax of 0 ″1. In short,

$$d = \frac{1}{p}$$

where *d* is measured in parsecs and *p* is measured in seconds of arc.

The star Proxima, mentioned above, has a parallax of 0 ″78. Using our formula, $1/p = d$, tells us that Proxima is 1/.78, or 1.28 pc from us. Thus light takes about 4.2 years to reach the earth from this star. Proxima is invisible to the naked eye but is a part of the multiple system known as Alpha Centauri. Alpha Centauri is in the southern part of the sky and cannot be seen from the continental United States.

252

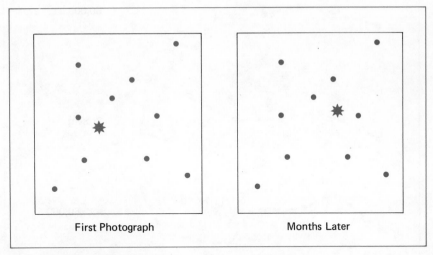

First Photograph     Months Later

Figure 9.2 **Parallactic Displacement of a Star**
The two diagrams represent photographs taken 6 months apart. The proper motion of
the star is taken into account before the parallactic displacement is measured.
Two photographs, however, are not sufficient to determine a reliable parallax. The
remote stars, shown as round dots, are too distant for their apparent positions
to be measurably changed by the earth's orbital motion.

Measuring the largest of all stellar parallaxes is about as difficult
as measuring the width of a quarter at a distance of about 2 miles.
Measuring smaller parallaxes is even harder. Generally, the minimum probable error of a well-determined direct trigonometric parallax is no less than 0″005, and in most cases it is probably larger.
At a distance of 1 pc, the probable error is 0.5 percent of the parallax. At 10 pc, it is 5 percent of the parallax, and at 100 pc, 50 percent. A distance of 100 pc is the practical limit of direct determinations of parallax.

There are less than 40 stars (counting multiple systems as single stars) within 5 pc (16 light years) of the earth, and 80 percent
of them are too faint to be seen with the naked eye. The distance of
Proxima, 1.28 pc, is a good approximation of the average separation
of stars in the sun's neighborhood. If 100 pc is the limit of direct
parallax measurements, and our galaxy is about 30 kpc in diameter,
then obviously other methods of determining distance are needed if

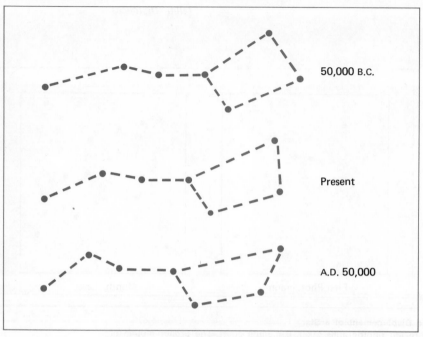

Figure 9.3 **Proper Motions in the Big Dipper**
Over a long period of time, the proper motions of the stars will alter the form of the familiar constellations. The picture shows the estimated change in the Big Dipper over a period of 100,000 years due to the proper motions of its component stars.

we are to understand the structure of our galaxy and the galaxies beyond it. These methods will be introduced later in the text. Fundamentally they are all based on information derived from direct parallax measurements and the length of the astronomical unit, which in turn depends on the speed of light. Refinement of this constant automatically refines our measures of stellar distances.

## 9.2 STELLAR MOTIONS

In 1718, Edmund Halley called attention to some bright stars that had moved from the positions given in ancient star maps and catalogs. However, the ancient measurements were not sufficiently precise to permit an accurate evaluation of this motion. There are several types of stellar motions. Most of the changes noticed by Halley were due to *proper motion*, the apparent annual change in the posi-

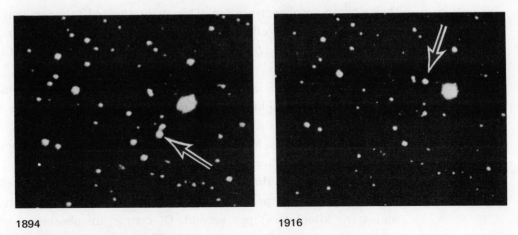

1894                                    1916

Figure 9.4  **Barnard's Proper Motion Star**
The star with the greatest known proper motion (10.3″ a year) moves a distance
equal to the diameter of the full moon in 180 years. Yerkes Observatory photograph,
University of Chicago.

tion of a star as seen from the earth when the effect of all other
variables has been removed. It is measured by comparing the present position of the star with its position a number of years before.
To achieve precise results, one needs an old but accurate star catalog, of which, fortunately, there are several. Comparing the former
position of the star with its present position, discounting the effect
of all variables such as parallax and precession, and dividing by the
number of years since the catalog was made gives an astronomer
the star's proper motion. Although the stars are moving rapidly
through space, they are so far away that their proper motions are
small. Over a long enough period of time, however, the proper motions of the stars will alter the familiar forms of the constellations,
as Figure 9.3 shows. On the average, nearby stars appear to have
moved more than distant ones, other things being equal. The star
with the largest proper motion is Barnard's star (Figure 9.4), which
moves approximately 10″ per year. This amounts to 0.5°, the diameter of the full moon, in 180 years.

Sometimes a group of stars, usually in a small region of the sky,

255

has a common proper motion. This is interpreted to mean that the stars are moving together; such a group of stars is referred to as a *moving*, or *galactic*, *cluster*. We shall study these clusters in Chapter 10.

Proper motion is a measure of apparent movement perpendicular to the line of sight. Stars in general move randomly in the sky, and each motion has a component parallel to the line of sight. This is the star's *radial velocity*. When a source of light approaches or recedes from the observer, we can measure the component of its velocity parallel to the line of sight by noting the Doppler effect (Section 3.8) in its spectrogram. This gives us the radial velocity in miles (or kilometers) per second. Of course, an observer on the earth is moving in the earth's orbit. This motion must be allowed for; hence radial velocities are given in terms of the sun's position rather than that of the earth.

Proper motion and radial velocities cannot be combined. Proper motion is given in seconds of arc per year, whereas radial velocities are given in miles per second of time. If no further information is available about the star, this is as far as one can go in analyzing its motion. However, if the astronomer knows, in addition to these two quantities, the distance of the star, then its proper motion in seconds of arc per year can be converted into its *tangential velocity* (Figure 9.5). This is its velocity in miles (or kilometers) per second perpendicular to the line of sight. If a star's tangential and radial velocity are known, we can compute its *space velocity*, or the speed of its movement through space with respect to the sun. From the observed direction of its proper motion, we can also compute its direction of movement. Space velocities in the vicinity of the sun average from 5 to 20 miles per second.

The space velocities of nearby stars are in part a reflection of the motion of the sun. In 1783, using an analysis of proper motions of 13 stars performed two centuries before, William Herschel arrived at the conclusion that the sun was moving in the general direction of the bright star Vega. The point toward which the sun is

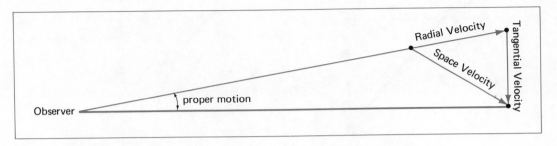

**Figure 9.5  Stellar Motion**
Proper motion is the amount of change in the position of a star during one year due
to its own motion with reference to the sun. Radial velocity along the line of sight in
miles or kilometers per second is determined from the Doppler effect, using a
spectroscope. If the distance of the star is known, its proper motion can be
translated into a tangential velocity perpendicular to the line of sight. When both
tangential and radial velocities are known, the space velocity, which is the resultant
of the two, can be computed.

moving with reference to the stars in our vicinity is called the *solar
apex* and is in the northern heavens in the constellation of Hercules.
The solar apex has not moved much since Herschel's determination
of its position.

The position of the solar apex can be determined in either of
two ways. First, one can do as Herschel did and consider the proper
motions of the stars. If the stars around us are moving at random
and we are moving through the region of space containing them,
the stars near the solar apex will appear to spread out as we approach them. Stars 90° removed from the solar apex in a great circle
perpendicular to our direction of motion will appear, on the average,
to drift away from the apex and toward an opposite point called the
*antapex* (Figure 9.6). Stars at the antapex will appear to draw together as we move from them. Solar motion can also be isolated by
an observation of radial velocities. On the average, stars near the
apex should show a preference for velocities of approach, and those
at the antapex should show a preference for velocities of recession.
Stars halfway between these two points should show no preference
whatever. The results of the two methods are consistent and yield
a velocity of about 12 miles per second for the sun. The solar sys-

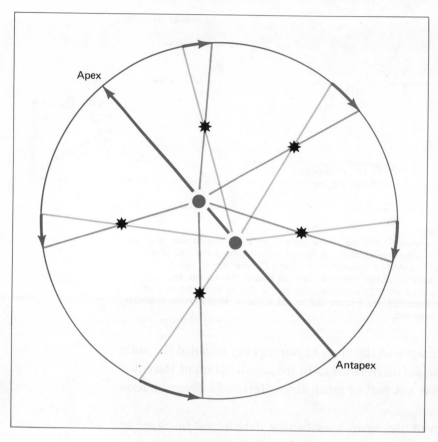

**Figure 9.6 Solar Motion**
As the sun moves away from the antapex toward the apex, the relative positions of
the nearby stars are affected. If these stars had no proper motion of their own,
their apparent motion would be a steady drift away from the apex and toward
the antapex as shown. The stars seem to drift away from the point in the sky toward
which the sun is moving. It was from observations of this type that William Herschel
first determined that the sun was moving among the nearby stars.

tem shares this motion. The earth goes around the sun with an average velocity of 18½ miles per second and shares the solar motion
of 12 miles per second. Consequently, the earth does not travel in a
closed ellipse with respect to the stars around the solar system, but
rather in a helix whose principal axis points toward the constellation of Hercules. The apex has a right ascension of 18$^h$0$^m$ and a declination of +30°. Although it might at first appear that we could use
the motion of the earth along with the sun due to solar motion to
obtain, eventually, a longer baseline for finding trigonometric par-

allaxes, this motion does not have the periodicity of the earth's orbital motion and does not yield a reliable parallax for single stars. For large numbers of stars, however, a *statistical parallax*, useful for certain purposes, may be derived. The sun's motion with respect to more distant stars is in the direction of the constellation of Cygnus; this motion is part of the rotation of the galaxy.

## 9.3 STELLAR RADIATION

The quality and the quantity of light received from a star can be used to analyze the characteristics of the star as an individual and the characteristics of stars in general. The principal instrument for analyzing the quality of starlight is the spectrograph. Spectral analyses of the sun, which is a typical star, were treated in Section 8.7. We shall now proceed to a study of stellar radiation from other sources.

After the introduction of the spectrograph to astronomy, spectrograms of many stars were made and a method of classifying stellar spectra was developed. Originally, the spectral classification system was alphabetical, but as more became known about the nature of the stars and of spectroscopy as well, departures from the previous alphabetical order were made. The majority of stellar spectra are now classified as one of seven types, O, B, A, F, G, K, or M, according to their temperature (Figure 9.7). The spectra of about 500,000 stars have been classified in this way. In addition, there are some stars whose spectra have peculiar features. These stars are sometimes classified as type W, R, N, or S. The digits following each letter classification indicate subdivisions. The hottest Class F stars, for example, are the F0's, and the coolest are the F9's. After the F9 stars come the slightly cooler G0 stars. A lower-case "e" denotes the presence of bright (emission) lines. The temperature and chief characteristics in the visual region of various types of stars are given below:

| Type | | Star |
|---|---|---|
| O6 | | λ Cephei |
| B3 | | η Aurigae |
| A0 | | δ Cygni |
| F2 | | β Cassiopeiae |
| G2 | | η Pegasi |
| K5 | | γ Draconis |
| M5 | | α Herculis |
| N0 | | 19 Piscium |
| Se | | R Geminorum |

Figure 9.7 **The Principal Types of Stellar Spectra**
Each spectrum shown is that of a typical member of its class. The bright lines on a
dark background above and below each spectrum are the comparison spectrum
used for calibration. Photograph from the Hale Observatories.

*Class O.* 30,000°K and hotter. These are blue giants. The spectrum shows ionized hydrogen (weak), helium, oxygen, and nitrogen. (See Section 9.4). Wolf-Rayet Class W stars are similar but show bright lines. Class O and W stars are the hottest stars.

*Class B.* 11,000° to 25,000°K. These are also blue giants. Neutral (un-ionized) helium is prominent early in this class but fades near the end. Some spectra show bright lines. Hydrogen lines grow stronger.

*Class A.* 7,500° to 11,000°K. These are usually white stars. Hydrogen is most prominent.

*Class F.* 6,000° to 7,500°K. Hydrogen begins to fade and ionized metals appear. The spectra are more complex. Neutral iron and chromium are present.

*Class G.* 5,000° to 6,000°K. These are yellow stars. A few molecules are present, and metals are more prominent.

*Class K*. 3,500° to 5,000°K. These are orange stars. Compounds are more prominent, and metals surpass hydrogen.
*Class M*. 3,500°K and cooler. These are red stars. Titanium and vanadium oxide bands are prominent, and neutral metals are strong.
*Classes R and N*. Carbon and carbon compounds are prominent. These stars are frequently called *carbon stars*.
*Class S*. Zirconium and lanthanum oxides are prominent.

We shall return to this stellar spectral sequence after we have considered the quantity of light received from the individual stars.

The word *magnitude* is used to describe the amount of light received from stellar sources. There are various types of magnitude scales. The first and simplest measures the *apparent magnitude* of a star, its brightness as seen from the earth. The ancients divided the visible stars into six classes. The brightest stars were considered stars of the first magnitude, and the faintest ones, just visible to the unaided eye, stars of the sixth magnitude. Modern astronomers have transformed this rather crude system into a more scientific magnitude scale. The average first-magnitude star is about 100 times as bright as a sixth-magnitude star. Thus a range of 5 magnitudes is equivalent to an increase in brightness by a factor of 100. Since $\sqrt[5]{100} = 2.512$, or approximately 2.5, a first-magnitude star is about $2\frac{1}{2}$ times as bright as a second-magnitude star; $6\frac{1}{3}$, or $(2.512)^2$, times as bright as a third-magnitude star; 16, or $(2.512)^3$, times as bright as a fourth-magnitude star; and 40, or $(2.512)^4$, times as bright as a fifth-magnitude star.

Since some objects are brighter than standard first-magnitude stars, and since telescopes see objects much fainter than those the eye can detect, the apparent-magnitude scale has been extended in both directions over the years. The brightest stars, some planets, the moon, and the sun have negative magnitudes. At the opposite extreme, stars photographed with the 200-inch telescope on Mt. Palomar have magnitudes as faint as +23.5. The sun, the brightest object in the sky, has a magnitude of −25.5. The difference in magnitude between the sun and the faintest star photographed with the 200-inch telescope is 50. Thus these stars differ in brightness from the sun by a factor of $100^{10}$, or $10^{20}$.

The apparent magnitude of the full moon is $-12.5$. We sometimes say, on a bright moonlit night, that "it looks as bright as day." Actually, this is very far from the truth. The difference between $-26.5$ and $-12.5$ magnitudes is 14 magnitudes, or $5 + 5 + 4$. Each 5 represents a factor of 100 in brightness, and the 4 represents a factor of about 40. Multiplying these numbers together, we find that the sun is approximately 400,000 times as bright as the full moon. For the night to equal the day in brightness, we would need more full moons than the hemisphere of the sky could contain.

If a star of moderate luminosity is close to us and a star of high luminosity is far away, the intrinsically fainter star may appear brighter. Hence apparent magnitude is, in part, an accident of the earth's position and is useless for studying the intrinsic characteristics of an individual star. However, if we also know the star's distance from us, we can calculate its *absolute magnitude*, the magnitude it would have if it were at a standard distance of 10 pc. The formula used is

$$M = m + 5 - 5 \log d$$

where $M$ is the star's absolute magnitude, $m$ is its apparent magnitude, and $d$ is its distance in parsecs. In studying the individual characteristics of the stars, a knowledge of their absolute magnitude is indispensable, since it removes from consideration the accidental circumstances of the star's position with respect to the earth.

In comparing absolute magnitudes, we are actually comparing the luminosity of the stars. Extremely luminous stars have negative absolute magnitudes of $-6$ or $-7$. The sun has an absolute magnitude of $+4.85$. In other words, if the sun were 10 pc from the earth, it would probably not even receive a Greek-letter designation. Table 5 in Appendix 1 gives the apparent magnitude for the nearest stars. If all the stars were at the same distance, as the ancients believed, their apparent brightnesses would be quite different from what we observe. For example, the star Sirius in Canis Major appears brighter by about 1.5 magnitudes than Rigel in Orion. Their absolute magnitudes, however, differ by about eight, and Sirius is fainter. Rigel has an absolute magnitude of −7.0 and Sirius of +1.41. Rigel is therefore about 1600 times more luminous.

Another factor also affects the apparent brightness of a star. This is the star's color. Since the human eye is more sensitive to red, orange, and yellow, and a photographic plate is more sensitive to violet and blue, the blue, hot stars appear brighter in photographs and red stars appear brighter to the eye. Hence both *photographic-* and *visual-magnitude* scales are needed. These scales are adjusted so that the visual and photographic magnitudes for an A0 star are equal. The difference between the two scales is called the star's *color index*. It is defined as the star's photographic magnitude minus its visual magnitude. It is a numerical measure of the star's color. Since hot blue stars are much brighter (have numerically smaller magnitudes) in photographs than they are visually, their color index is negative. Red stars, on the other hand, have a positive color index. Sometimes, differences in color are due to absorbing material between the star and the observer. When this occurs, the difference between the color index as determined from an examination of the star's spectral class and from observations of its absolute and apparent magnitude is called its *color excess*. The color excess may be used to estimate the amount of absorbing material between the star and the earth.

Two characteristics of stars, their spectral class and their absolute magnitude, can be plotted against one another on a graph to obtain what is called a *Hertzsprung-Russell* (or H-R) diagram (Figure 9.8). Such a diagram contains, not a random scattering of points, but a systematic arrangement of stars. The principal group of stars, running from the upper left to the lower right of the diagram, is called the *main sequence*. The sun, a G2 star with an absolute magnitude of +4.85, is in the main sequence. Stars above the main se-

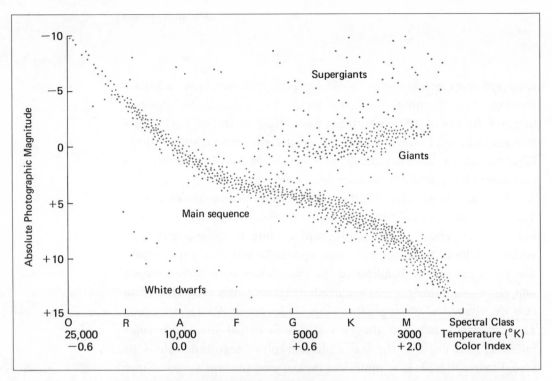

Figure 9.8 **The Hertzsprung-Russell Diagram**
Plotting the absolute magnitude of a star against its spectral class (or temperature) produces a point on a graph. Plotting the absolute magnitudes of all the stars whose distances are known against their observed spectral class produces this diagram. The vast majority of stars lie in the main sequence. The numerical importance of giants and supergiants is overemphasized because they are visible at greater distances than ordinary stars.

quence are brighter than stars of the same spectral class in the main sequence. These are *giant* or *supergiant* stars. To the left of and below the main sequence are a few stars of high temperature but low luminosity, and consequently small dimensions. These are *white dwarfs*. The *H-R diagram* is one of the most useful tools we have for studying the stars.

## 9.4 STELLAR ATMOSPHERES

In Section 3.8, we discussed the principle of the spectrograph and Kirchhoff's laws of spectra. The relevance of these laws and the

264

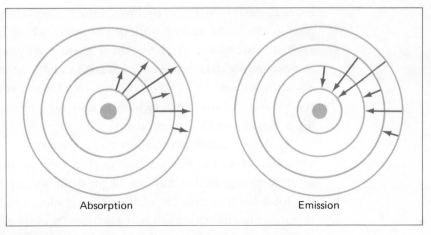

Figure 9.9 **Electron Transitions**
An atom can absorb light of a particular wavelength when an electron jumps from a lower orbit to a higher orbit (left). This causes dark lines (absorption lines) in its spectrum. When electrons jump from higher orbits to empty lower orbits (right), light is emitted. This produces bright lines (emission lines) in its spectrum.

laws of thermodynamics to analyses of solar energy was covered in Section 8.4. We turn now to a study of stellar spectra and what they tell us about the atmosphere of the stars.

The photosphere of a star is a dense gas capable of producing all wavelengths of visible light. This radiation produces the continuous background typical of nearly all stellar spectra. Between the photosphere and an earthly observer is the atmosphere of the star. Since it is tenuous and has a lower temperature than the photosphere, it produces dark lines in accordance with Kirchhoff's third law. The particular dark lines produced, as well as their relative strength and other characteristics, are explained by the behavior of atomic particles in the tenuous stellar atmosphere.

The conventional, and still useful, picture of the *atom* deduced by Niels Bohr (1885–1962) consists of a *nucleus* and a system of one or more orbiting *electrons*. The nucleus possesses practically all the mass of the atom, and the electrons revolve in specific orbits (Figure 9.9). If the atom absorbs a photon with the proper amount of energy, the electrons can move into higher (larger) orbits. The photon raises the *energy state* of the atom, which is then said to be *excited*. Absorption of energy of a certain wavelength produces a corresponding dark line in the spectrum of the atom. Reemission of a photon causes the electron to drop to a lower orbit and lowers the energy state of the atom (which is now no longer excited), producing a bright line in the spectrum.

If the absorbed photon has enough energy, an electron may be separated from the atom, which is then *ionized*. Since the net total positive charge of the nucleus of an ordinary atom is balanced by the total negative charges of the orbiting electrons, an ionized atom is positively charged. Each lost electron (and its attendant negative charge) represents an unbalanced positive charge on the nucleus. Atoms in this state are indicated by one or more plus signs following the chemical symbol of the element. For example, "$Fe^{2+}$" indicates doubly ionized iron. In more complex atoms containing many electrons, the process of ionization may continue through various steps. Each absorption of energy by the atom depends on the state the atom is in, the state at which it finally arrives, and what kind of an atom it is.

Normally, an atom remains in an excited state only for something like 100-millionth of a second. Since returning to a lower energy state releases light of the same wavelength as that absorbed, one might wonder why there are dark lines. In the atmosphere of a star, energy is reemitted about as rapidly as it is absorbed, so that seemingly the two actions should cancel out. However, the direction of the original emission from the photosphere in the continuous spectrum is along the line of sight. The direction of reemissions, on the other hand, is arbitrary, and the observer on earth receives only a small percentage of the original light. This produces the apparently dark lines of the spectrum. As we noted in our study of the spectroheliograph, the "dark" lines are not completely dark, although they appear so against the bright continuous background.

The ability of atoms to absorb and reemit light of particular wavelengths varies. The temperature of the star is the principal factor determining what atoms, at what energy level, are most productive of dark lines. In other words, the temperature of the star is mainly responsible for its spectral class, and consequently for its color, which is expressed numerically by its color index. Thus the H-R diagram can, for some purposes, be calibrated using the stars' spectral class, their effective temperature, or their color index.

Let us consider the significance of the main sequence. This is the diagonal group in the H-R diagram running from the extremely luminous Class O stars in the upper left to the low-luminosity Class M stars at the lower right. If all stars were on this sequence, we could, by examining a star's spectrum and placing it on the horizontal scale, running upward to the main sequece, and moving left from the intersection point to the vertical axis, find the absolute magnitude of the star. Substituting this value in the formula $M = m + 5 - 5 \log d$, which relates absolute magnitude, apparent magnitude, and distance, we could then determine the distance to the star.

However, the main sequence is not actually a line, but a belt about 1 magnitude in width; hence, an error factor of about 2.5 would be inevitable in determining the luminosity, and therefore the distance, of the star. Nor is the basic assumption of this method accurate. Not all stars of a particular spectral type are in the main sequence. The sun, for example, which is a Class G star, has an absolute magnitude of about +5, but there are Class G supergiants with absolute magnitudes of −6 or −7. Assuming that one of these supergiants was on the main sequence would mean an error of more than 10 magnitudes in our value for the luminosity of the star, and the resulting value for its distance would be grossly in error.

Fortunately, a giant or supergiant star is not simply a larger copy of a main-sequence star of the same type. There are inherent physical differences between the spectra of the giant stars and main-sequence stars that make it possible for us to refine the method of determining stellar distances proposed above. We can divide the stars in a particular spectral class into five subclasses of luminosity. This division is more or less vertical, in contrast with the horizontally plotted spectral classes in the H-R diagram. The subdivision is indicated by a Roman numeral following the code for the star's spectral class. The Roman numeral I is used to indicate supergiants, whose absolute magnitude ranges from −5 to −7. A small letter *a* is sometimes used after the I to indicate the brighter members of this subdivision, and a small *b* to indicate fainter stars. Type II stars are bright giants with absolute magnitudes of about −2; Type III stars are ordinary giants with magnitudes around 0; Type IV stars are subgiants with magnitudes of about +2; and Type V stars are main-sequence stars such as the sun. In the H-R diagram, the first four types are to the right of and above the main sequence (Figure 9.10). To the left of and below the main sequence are the white dwarfs, which are not given a luminosity classification. According to this system, the sun's spectral class is G2V.

Because excitation and ionization occur at a lower temperature in a less dense medium, the tenuous giant stars are cooler than main-sequence stars of the same spectral class. For example, the star Antares (α Scorpii) is about 220 times the diameter of the sun, but has only 10 times the sun's mass. Since its volume is approximately 10 million times that of the sun, its density must be about one-millionth the average density of the sun. In other words, an average cubic foot of the star Antares would contain one-millionth the mass of an average cubic foot of the sun. A star of such low density can produce the same *general* pattern of lines at a lower temperature than a corresponding main-sequence star. The lower-density star does not, however, produce *exactly* the same pattern of lines, or exactly the same intensity ratio of selected pairs of lines, as a main-sequence star in the same spectral class. For example, the ratio of the strength of an iron line to that of a strontium line (Figure 9.11) varies for stars in Classes I through V. By using stars whose distances are already known and whose abso-

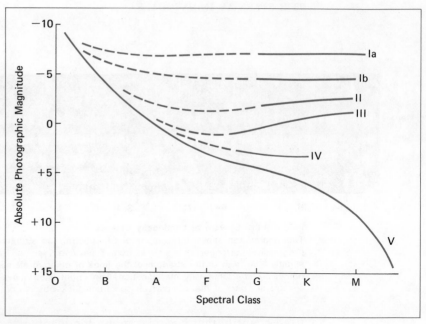

Figure 9.10 **Luminosity Classes**
This graph shows the absolute magnitudes of stars in different luminosity classes on or above the main sequence of the H-R diagram. A star's luminosity is determined from its spectrum. (See Figure 9.11.)

lute luminosities can be independently computed, we can set up empirical criteria for determining the luminosity of an observed star in a particular spectral class.

We determine the luminosity of a star by comparing the relative strength of particular pairs of lines in the spectra of stars in the same spectral class. A curve is drawn showing this variation, giving the strength of one line in terms of the other. Sets of these curves are made for each spectral class. When a spectrum is made, the densities of appropriate lines are measured and the ratios computed. From this information and the curves, the luminosity class, and hence the absolute magnitude, of the star is determined. The apparent magnitude is easily observed, so that its parallax, and thus its distance, can be computed. Parallaxes obtained by this method are called *spectroscopic parallaxes*. They are determined by an empirical process that works for stars which fit the pattern. The probable error is approximately 15 percent. Unlike the probable error of a direct trigonometric parallax, which is a

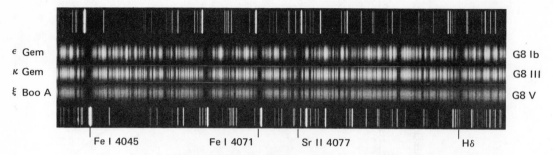

ε Gem — G8 Ib

κ Gem — G8 III

ξ Boo A — G8 V

Fe I 4045    Fe I 4071    Sr II 4077    Hδ

Figure 9.11 **Spectra of Luminosity Classes**
This photograph shows three types of G8 spectra. The strontium line (Sr II 4077) gets progressively stronger as we move from Type V to Type Ib. Iron (Fe) changes only slightly. The ratio of the strength of the lines of different elements is used to obtain a star's luminosity classification. The bright lines above and below the G8 spectra are part of the comparison spectrum. Lick Observatory photograph.

fixed quantity that becomes more dominant as the parallax becomes smaller, the probable error here is a fixed percentage of the parallax itself, and its absolute value diminishes with the parallax. Since the probable error of a direct parallax at 100 pc is about 50 percent, spectroscopic parallaxes are more reliable at this distance.

Since giant stars of a particular spectral class have nearly, although not exactly, the same temperature, and at the same time are hundreds or thousands of times brighter than those of the same class on the main sequence, they must, as a group, be much larger. Stellar radii can be determined using the Stefan-Boltzmann law (Section 8.4). Suppose that the spectrum of a star indicates a temperature of about 4000° K, two-thirds that of the sun, and its luminosity, $L$ (its total emitted energy), is 80 times that of the sun. Energy per unit of area is proportional to $T^4$; the area is proportional to $R^2$. ($R$ is the radius of the star in terms of the radius of the sun). $L$ is then proportional to $R^2T^4$ and

$$R^2 = \frac{L}{T^4}$$

where $R$, $L$, and $T$ for the sun are taken as units. $R^2 = 80/(\frac{2}{3})^4$, or 405, and $R$ is approximately 20 solar radii.

Some stellar radii have been measured with the interferometer

270

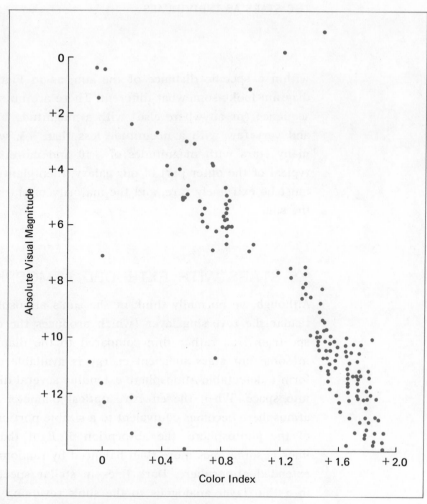

Figure 9.12 **An H-R Diagram for Stars Within 10 Parsecs of the Sun**
If a sphere 10 parsecs in radius with the sun at the center is fairly typical of our
region of space, then giant and supergiant stars must be extremely rare. No star with
100 times the luminosity of the sun has been found within this distance, whereas
many stars $\frac{1}{100}$ as bright, or even fainter, are known. The stars in the lower left
are white dwarfs.

(Section 3.10) and some have been inferred from observations of
eclipsing binaries. These findings are consistent with the results
obtained by the method explained above.

Giant and supergiant stars often seem more numerous than
they really are. In Figure 9.8, all the stars for which information
is available are plotted. If, however, we limit ourselves to stars

within a specific distance of the sun, as in Figure 9.12, the H-R diagram looks somewhat different. There are no stars on the main sequence (or anywhere else) with a magnitude brighter than zero and very few with a magnitude less than +3, whereas there are many stars with magnitudes of +10 and more. If this region is typical of the outer part of our galaxy, then giants and supergiants must be extremely rare, and the majority of stars are fainter than the sun.

## 9.5 STARS WITH EXTENDED ATMOSPHERES

Although we normally think of the sun's atmosphere, and in particular the reversing layer (which produces the dark lines of the spectrum), as rather thin compared to the diameter of the sun, in some hot stars sufficient energy is available at the surface to form a detectable atmosphere extending several diameters outward into space. When the effective optical diameter of the reflecting atmosphere becomes equivalent to a sizable portion of the diameter of the photosphere, the absorption of light that produces dark lines is sometimes more than balanced by random reflection in the extended atmosphere. Dark lines in stellar spectra are produced in a thin layer analogous to the sun's reversing layer, but bright reemission lines can come from an extended atmosphere appreciably larger in diameter than the star itself. The spectra of 10 percent of the hot stars in Classes B0 to B3 have emission lines. These stars frequently rotate rapidly about axes, which, when the axes are perpendicular or nearly perpendicular to the line of sight, produce broadened lines in the spectra.

Another type of star with an extended atmosphere is the *shell star* (Figure 9.13). This is a star surrounded by a shell of gas, whose spectrum characteristically shows narrow dark lines (that is, lines not broadened by rotation). P Cygni stars and Wolf-Rayet stars also have very extended atmospheres. The Wolf-Rayet stars,

272

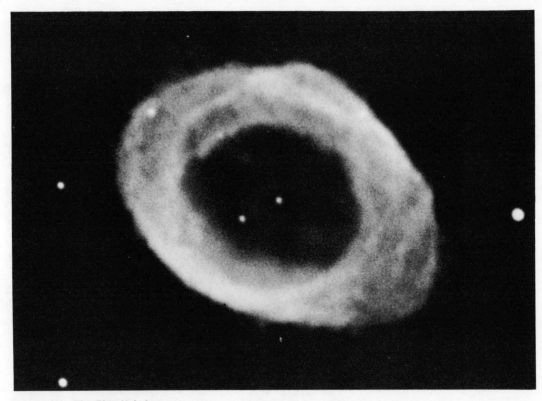

Figure 9.13 **The Ring Nebula**
This planetary nebula in the constellation of Lyra is actually an expanding shell of
gas around a very hot star. Photograph from the Hale Observatories.

which have been studied for over a century, are among the brightest
stars known. They have surface temperatures of 50,000°K and
broad bright lines in their spectra, indicating rotational velocities
as great as 2000 miles per second.

Another type of celestial phenomenon that, in a sense, constitutes an extended atmosphere for a star is the *planetary nebula,*
of which about 400 are known and many more must certainly
exist in our galaxy. They are called planetary nebulas not because
they are associated with planets, but because some of them appear
in the telescope as small greenish disks reminiscent of the appearance of Uranus and Neptune. Their diameters are about 20,000
AU's. The central star of a planetary nebula has a mass approximately equal to that of the sun, but a much smaller diameter;
consequently, it has a higher density than the sun. These are hot

273

stars, approaching the Wolf-Rayet stars in temperature, and they sometimes show the typical Wolf-Rayet spectrum. Most of their radiation is in the ultraviolet range, but reflection and rereflection within the shell of gas that forms the nebula degrades it to visible radiation. The Doppler shift of the planetary nebulas shows that they are expanding at velocities in the neighborhood of 12 to 15 miles per second. They probably have a relatively short life and may represent a stage in the evolution of particular types of stars. The mass of the nebula is surprisingly high, about one-tenth the mass of the central star. This indicates that a planetary nebula is not a remnant of a nova, or "new star."

## 9.6 STELLAR INTERIORS

A detailed analysis of the mathematics and physics needed to understand the interplay of forces in stellar interiors is beyond the scope of this text. However, the interior is obviously the source of stellar energy, and we shall investigate where this energy comes from, using the sun as a typical example. The Einsteinian equation $E = mc^2$ shows the energy output of a star ($E$) to be a function of the relationship between the amount of mass undergoing conversion ($m$) and the velocity of light ($c$). The energy is measured in ergs, the mass in grams, and the velocity of light in centimeters per second. The type of matter (that is, its chemical composition) does not affect the equation or the final result.

Two basic processes are responsible for the release of nuclear energy. The first is *fission* of heavy, complex atoms, such as uranium and plutonium; this process was the source of the energy released by primitive atomic bombs. The second process is one of *fusion* and involves only the lighter atoms. It is this process that is responsible for the action of the hydrogen bomb. In the sun, as we noted in Chapter 8, four hydrogen atoms are combined to make one helium atom. Since four hydrogen atoms have a slightly greater mass than one helium atom, the excess (about 0.7 percent of the mass of the four hydrogen atoms) is converted into energy. That is, 1 gram of hydrogen is converted into 0.9929 gram of helium; the remaining 0.0071 gram becomes energy. Since $E = mc^2$, the conversion of 1 gram of hydrogen each second into approximately 1 gram of helium releases 640 million killowatts of energy.

Like the sun, stars are believed to contract from large tenuous clouds of gas as a result of their own gravitational attraction. A contracting gas becomes hot, and when the central temperature of a star such as the sun reaches 10 to 15 million degrees, the fusion process starts. There are actually two processes. In stars of the sun's class and cooler stars, fusion is accomplished by what is called a *proton-proton reaction*. Stars hotter than the sun also transform hydrogen into helium to release energy, but they do so primarily by a somewhat different process, called the *carbon cycle*, since atoms of carbon, nitrogen, and oxygen are involved.

The rate at which a star lives and uses its fuel depends on its mass and increases enormously with it. For example, a star with a mass three times that of the sun lives about 120 times as fast and lasts one-fortieth as long, or about 250 million years (a short time, astronomically speaking). On the other hand, a star with one-half the mass of the sun lives about one-fifth as fast and lasts two and a half times as long as we expect the sun to live.

## 9.7 INTRINSICALLY VARIABLE STARS

Throughout our discussion of stellar energy, we have ignored the fact that the quantity and quality of radiation emitted by many stars varies. Intrinsic variation in starlight is almost always connected with a change in the diameter and temperature of the star. Moreover, variable stars are sufficiently numerous for astron-

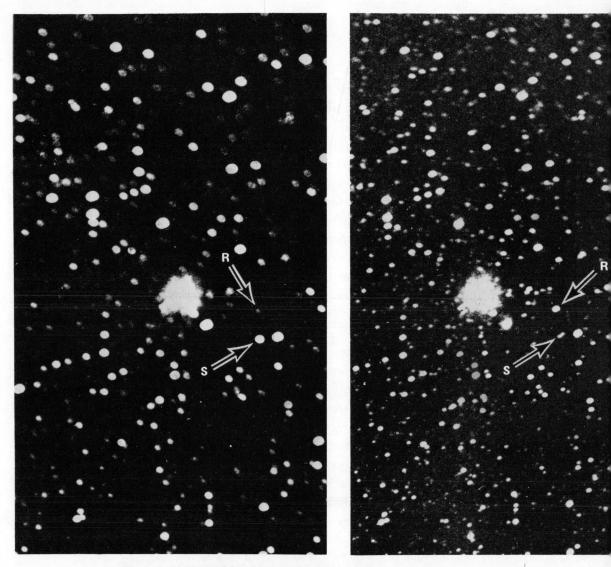

Figure 9.14  **The Variable Stars R and S Cassiopeia**
These photographs show the change in brightness of two variable stars with respect
to stars of constant brightness. Yerkes Observatory photograph, University of Chicago.

omers to feel that they may represent a normal stage in the
evolution of at least some, and perhaps all, stars.

There is a wide variation in the nature of intrinsic stellar
variability. When a variable star (Figure 9.14) is observed and its
brightness plotted against elapsed time, a *light curve* indicating

276

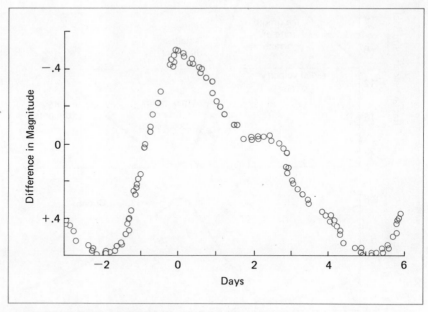

**Figure 9.15  The Light Curve of a Cepheid Variable Star**
In a Cepheid variable, the rise to maximum light is abrupt. The fall is less so, and there is sometimes a slight hump about halfway down. The time between similar phases is the star's period, in this case about 7 days. The star is Eta Aquilae, a classical Type I Cepheid.

its period is obtained (Figure 9.15). The curve rises as the star brightens and falls as it grows dimmer. The distinctive character of any variable star is generally indicated by its period.

The shape of the light curve is also important in distinguishing different types of stars that may have the same period. *RR Lyrae stars,* named for the star which is their prototype, have periods less than or equal to a day and an average absolute magnitude of between zero and +1. About 500 or more are known in the Milky Way. Since they are found in globular clusters, they were originally called *short-period,* or *cluster-type,* variables. The classical Cepheid variable stars, named after their prototype, Delta Cephei, have periods ranging from 1 to 50 days. These stars are now referred to as *Type I Cepheids. W. Virginis,* or *Type II, Cepheid*

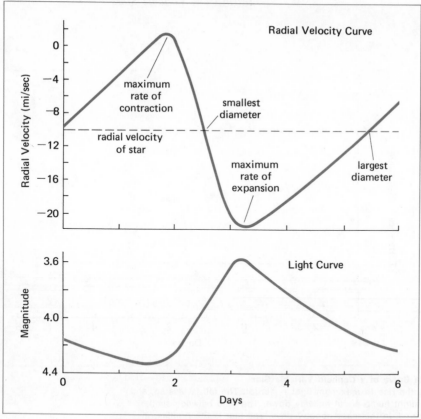

**Figure 9.16  Light and Velocity Curves of a Classical Cepheid**
The change in the apparent velocity of the star is due not to the motion of the star itself, but to the pulsation as the surface, which expands and contracts, moves toward and away from the observer. Notice the correlation between the rate of expansion and contraction of the star and the light curve.

*variables* have periods of from 12 to 40 days but have a slightly different light curve from classical Cepheids. They are about 1.5 magnitudes fainter than classical Cepheids of the same period.

At the time that a light curve is being obtained photometrically, spectroscopic observations of the star may be made. These show that, in step with the variation in light, there is a variation in the Doppler shift, indicating a change in the velocity of the surface of the star with respect to the earth (Figure 9.16). In addition, there is a change in the spectral class of the star. The greatest negative (approach) velocity is recorded at the time of maximum brightness when the star is at its hottest. As the light curve drops, the expansion of the star slows down and another contraction begins. When the light reaches a minimum, the surface of the star is moving most rapidly away from the earth (contracting) and is at its coolest. As the light increases, the star stops contracting and begins to expand again. This indicates that the variation in light is connected with pulsations or changes in the star's diameter and changes in its spectral class. However, the relationship is not simple, and various levels of the star do not pulsate in unison. The Doppler shifts are produced in the star's atmosphere, but the change in the quantity of radiation has its source in the star's interior. In 1938, Martin Schwartzschild suggested that the interior pulsates independently of the outer layers, which do not all vibrate together. The light reaches a maximum when the inner layers are most compressed but lags behind the smallest volume of the outer layers and comes when they are expanding fastest.

In 1912, Henrietta Leavitt discovered a relationship between the apparent magnitude and the period of variable stars in the smaller Magellanic cloud (one of a pair of nearby galaxies). The distances between galaxies are so great that the distance of all the stars in the Magellanic clouds may be regarded as essentially the same and their apparent magnitude as differing from their absolute magnitude by a constant amount. Hence a relationship between their apparent magnitude and their period may be taken as a relationship between their absolute magnitude and their period as well. Unfortunately, at that time, the distance (and hence the absolute magnitude) of the Magellanic clouds was not known and the curve for the period-luminosity relationship (Figure 9.17) could not be calibrated. Eventually, however, a statistical parallax for the clouds was obtained.

Leavitt's period-luminosity relationship has proved an invaluable yardstick for measuring the distances of Cepheid variables in our own galaxy and other galaxies as well. Their period and their apparent magnitude can, like the period and apparent magnitude of stars in the Magellanic clouds, be determined from ob-

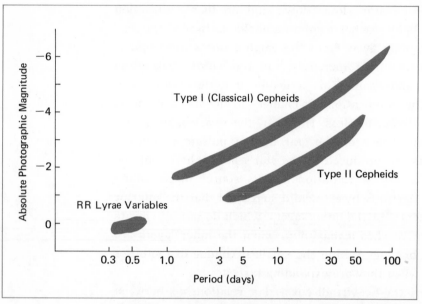

**Figure 9.17  The Period-Luminosity Relationship in Cepheids**
When the period of light variation of a Cepheid is known and the type of Cepheid is established, its average absolute magnitude may be read from the graph, and its distance determined. Note, however, that some uncertainty is introduced because the relationship between the star's period and luminosity cannot be shown by a single line but only as a region about 1 magnitude in width.

servation. Their absolute magnitude can be obtained from the period-luminosity relationship, and their distance can be calculated from the formula for finding the absolute magnitude when the apparent magnitude and distance are known ($M = m + 5 - 5 \log d$). In this case, of course, $M$ is known and $d$ is the dependent variable.

Cepheid variable stars are supergiants and are visible for very long distances. This, plus the fact that they exist in many groups of stars, enhances their value as yardsticks for determining stellar distances. However, all determinations of distance involving observed luminosities contain some degree of uncertainty. The inverse-square law regarding the decrease in electromagnetic radiation through space depends on absolute clarity of transmission. Between the stars there is unquestionably, in certain regions, absorbing material. We shall consider the overall problem of the detection of interstellar material and its effects in Chapter 11.

In addition to the problems caused by interstellar absorption, parallax determinations from Cepheid and RR Lyrae variables suffer because the period-luminosity relation is graphically equivalent not to a line but to a region about 1 magnitude wide. This introduces a corresponding error in value obtained for distances. Another source of error in the early days of investigation was the assumption that there was only one type of Cepheid with a period longer than a day. Early in 1950, however, Walter Baade found that, although he could photograph many Cepheids in the Andromeda galaxy, the RR Lyrae stars were not bright enough to be seen with a 100-inch telescope. He calculated that they should be easily visible once the 200-inch telescope went into operation, but this was not so. The stars Baade had been photographing were about 1.5 to 2 magnitudes brighter than anyone had suspected. It was then determined that there were two types of Cepheid variables, as in Figure 9.17. As a result of this discovery, estimates of the distances of galaxies outside the Milky Way were doubled. Cepheid variable stars continue to be a valuable tool for determining the distances of the nearer extragalactic systems and provide a steppingstone to the establishment of a scale for the distance of remote galaxies.

Another type of variable star is the *long-period variable* (Figure 9.18). These stars are not as homogeneous as the shorter-period variables with periods of less than 50 days. There are various types of long-period variables, and they are the most common among the pulsating stars. Basically, they are red giants and supergiants. The star Mira is a good example of this group. Their temperature is low, between 2000°K and 2500°K, and their variability does not have the precision of the short-period variables. Generally, the longer the mean period is, the less regular the star will be. Indeed, many red variable stars show no sign of regularity.

The most spectacular of all variable stars is the nova (Figure

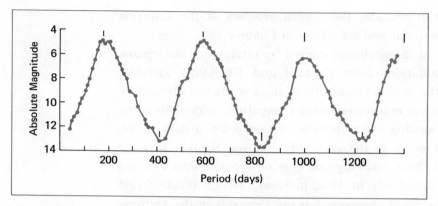

Figure 9.18 **The Light Curve of Chi Cygni**
This diagram shows the light curve of a long-period variable of the Mira type, determined by the American Association of Variable Star Observers.

9.19.) The word *nova* means new, but the stars to which this term is applied are not new. They simply increase thousands of times in brilliance in a short period. They are explosive, or eruptive, stars. Novas have been known from ancient times and were recorded by Chinese astronomers thousands of years ago. Often, when a nova appears, we can go back to old photographs of the region of the sky in which it is located and find a record of the star before it exploded. There are an estimated 30 or 40 nova outbursts in our galaxy per year, and they may be a normal evolutionary process in certain types of stars. Ordinary novas increase as much as 10,000 or more times in brightness in a short period, then fade rather slowly, taking perhaps as long as a year to return to approximately their former state (Figure 9.20). Apparently, these outbursts are only "skin deep." Expanding shells of gas around novas are common. However, they are transitory and do not have nearly the mass or the size of planetary nebulas. The mass of a planetary nebula is about one-tenth that of the central star, whereas the mass of a nova envelope is probably less than 1/10,000 that of the central star. We do not know precisely what causes a nova.

Figure 9.19  **Nova Herculis**
These two views graphically show the increase in brightness of the star between
March 10 and May 6, 1935. Lick Observatory photographs.

283

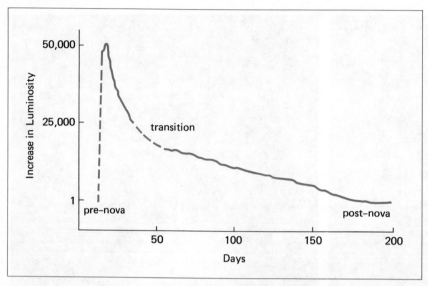

**Figure 9.20  A Light Curve for a Typical Nova**
The rise to maximum brilliance is abrupt, and the original magnitude of the star must be determined by photographs taken of the area before the outburst. The outburst represents an increase of tens, even hundreds, of thousands of times the star's original intrinsic brilliance. The drop back to approximately the original magnitude of the star is slow and may be accompanied for a while by minor fluctuations.

The most spectacular type of nova is the *supernova*. This is one of the most spectacular of natural events and probably represents the nearly complete destruction of a star. A supernova may flare up to billions of times its former brightness and may reach an absolute magnitude of −20. Supernovas were recorded in 1054, in 1572, and in 1604. Since they are among the brightest natural objects in the sky, they can be observed even in exterior galaxies, beyond the limit at which ordinary, or ever supergiant, stars are visible. The light curve of a supernova is similar to that of an ordinary nova, but much brighter.

The Crab nebula in the constellation of Taurus is all that remains of the supernova of 1054. It is a strong source of radio waves. Inside the nebula, barely detectable, is a small white star that has been identified as a pulsar.

284

## 9.8 STELLAR ROTATION AND MAGNETIC FIELDS

If a star rotates on an axis nearly perpendicular to the line of sight, one limb approaches the earth and the other recedes, while the central part of the disk does neither. The resulting Doppler shifts broaden the lines of the spectrum. If the star does not rotate, or if its axis is more or less parallel to the line of sight, the lines will be sharp and narrow. The profile of the photographic density in the spectrogram of lines broadened by rotation is characteristic and makes it possible to distinguish this type of broadening from others. Nonmultiple blue main-sequence stars seem to have the highest rotational speeds. Cooler stars have more moderate speeds, and this has been interpreted as indicative of the possible existence of planetary systems about them. Giant stars have slower rotational speeds than corresponding main-sequence stars.

Like sunspot spectra (Section 8.5), the spectra of some stars show strong magnetic fields. H. W. Babcock, in 1958, listed 89 stars with definite magnetic fields, and fields in excess of 5000 G have been detected.

# QUESTIONS

(1) How are the distances of the nearer stars determined?

(Section 9.1)

(2) What is a parsec? (Section 9.1)

(3) What is meant by the proper motion of a star?

(Section 9.2)

(4) What must one know in order to determine the space velocity of a star? (Section 9.2)

(5) Describe the spectral classification of stars. (Section 9.3)

(6) What is the difference between the apparent and absolute magnitude of a star?                                     (Section 9.3)

(7) Draw an H-R diagram, and indicate location of the main sequence.                                              (Section 9.3)

(8) How does the Bohr model of the atom explain absorption lines and emission lines in spectra?                   (Section 9.4)

(9) How does an astronomer determine the luminosity class of a star?                                               (Section 9.4)

(10) What is a planetary nebula?                       (Section 9.5)

(11) How is atomic fusion related to the energy of the sun?
                                                       (Section 9.6)

(12) What is a classical Cepheid?                     (Section 9.7)

(13) How does the period-luminosity relationship of a Cepheid help us to determine its distance?                  (Section 9.7)

(14) What is the difference between a nova and a supernova?
                                                       (Section 9.7)

# CHAPTER 10
# GROUPS
# OF STARS

Up to now, we have spoken of the stars as if they were single, isolated objects. However, multiple systems of stars are well known. Within 5 pc of the sun, there are at least 11 double (or binary) stars and at least 3 triple stars. Of the 54 individual stars in this vicinity, then, 31, or more than half, are part of multiple systems. At greater distances, double stars become increasingly difficult to detect. However, some 40,000 binaries are known; and although they may be a minority of the total stellar population, they are certainly a significant one.

## 10.1 VISUAL BINARY STARS

When the heavens are examined with a small telescope, what had appeared to the naked eye as a single star occasionally turns out to be two stars. A good example of this is the star Castor in the constellation of Gemini. Sometimes, what appears to be a double star is merely an accidental alignment of two stars in nearly the same direction (but separated by considerable distance) along the line of sight. An example of this is the star Deneb in Cygnus, which appears to be a double star but is actually two separate stars, one about twice as far away as the other. However, the majority of apparently close stars are members of double or other multiple systems. Such stars are called *visual binaries*. Being close together, they revolve about their common center of mass. Just as the path of a satellite enables us to measure the mass of a planet using Kepler's laws, so too the motions of binary stars permit us to measure their respective masses.

Visual binary stars are close enough to the earth and separated from each other by a sufficient distance so that they may be observed either visually or photographically with existing telescopes. The detection of such a star depends on the apparent separation of the two stars as seen from the earth and the resolving power of the optical system.

Visual binaries are usually detected by observing the mutual revolution of the stars about their common center of gravity (Figure 10.1). In some instances, however, they are so far apart that their orbital motion is extremely slow, and centuries or millennia may pass before this motion can be detected. When two relatively close stars share a common proper motion and have the same parallax, they are assumed to have a slow orbital motion. This provides a second method of detecting binaries. The first visual binary to be detected was Mizar, which was separated in the telescope in 1650; proof of orbital motion was first obtained for Castor, by William Herschel in 1804.

The orbital motion of a pair of stars can be measured photographically or visually. Two quantities are needed in order to obtain

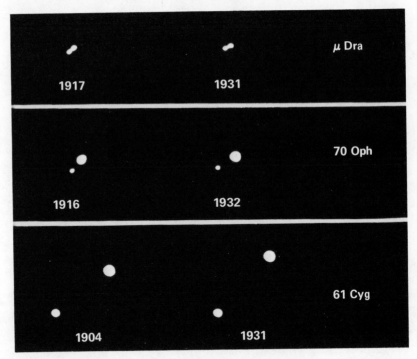

**Figure 10.1  Orbital Motion in Binary Systems**
These three binary stars were photographed at intervals of 14, 16, and 27 years.
The change in the orientation of each pair is evidence of their orbital motions.
Yerkes Observatory photographs, University of Chicago.

an observation. One is the separation, in seconds of arc, between those two stars measured along a line joining them. The other is the position angle, usually of the fainter star with respect to the brighter one. This angle is measured eastward from the north, from 0° to 360°, with the brighter component at the center and the line joining it to the fainter one the terminus of the angle measured from the true north point. Of course, the minimum observable separation of a pair of stars depends on the size of the telescope used. In the 82-inch telescope at McDonald Observatory, for example, the minimum observable separation is 0 .″ 1.

When enough high-quality observations have been accumulated, the *apparent relative orbit* (Figure 10.2) of the system can be computed. This orbit will be an ellipse that shows the motion of the fainter companion with respect to the bright primary. If the orbital plane is perpendicular to the line of sight, this will also be the

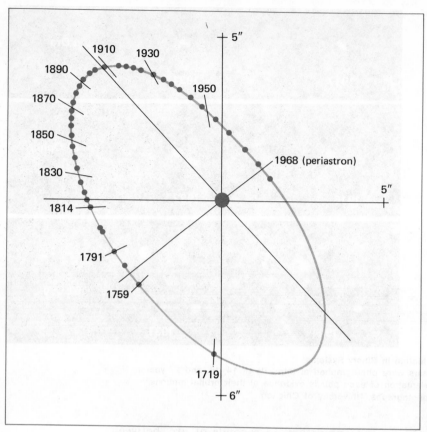

**Figure 10.2  The Apparent Relative Orbit of Castor**
The period of revolution of this double star is 380 years. It has been observed for over 2½ centuries. Here, the orbit is plotted as if the primary star were motionless and its companion were revolving around it. In fact, both are revolving about a common center of mass. In addition, the true orbit is projected on the plane of the sky perpendicular to the line of sight. (See Figure 10.3.) The two stars passed periastron (the point at which they are closest to each other) in 1968. Note that the primary star in the projection does not appear at the focus of the elliptical orbit of its companion. (Orbit determined by K. A. Strand).

*true relative orbit* (Figure 10.3); but in most cases there is an angle of inclination between the orbital plane and the plane perpendicular to the line of sight, which is usually called the *plane of the sky*. There are geometric methods of obtaining the angle of inclination of the orbital plane of the sky for a visual binary system. If this is done, the foreshortening of the dimensions of the orbit immediately becomes evident and the true orbit can be traced. This orbit is still a relative orbit because position angles and separations

290

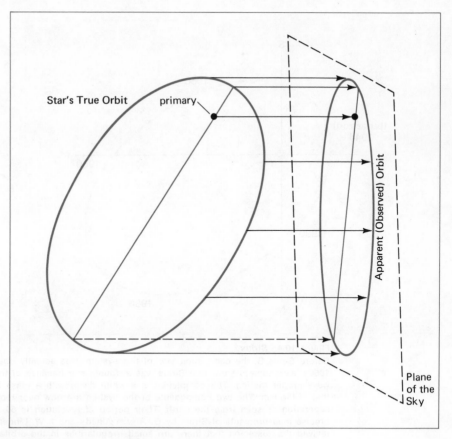

Figure 10.3 **Apparent and True Relative Orbits**
Practically no binary orbit is exactly in the plane of the sky. However, this fact is
ignored when the original observations are made and the apparent orbit is
calculated. We can, by certain mathematical operations, evaluate the inclination of
the orbital plane of a visual binary to the plane of the sky. When the apparent orbit
has been rotated into the position of the true orbit, the primary star takes a position
at the focus of the relative orbit of its companion.

are measured relative to the brighter of the two components of the
system. If the positions of the stars in the system are measured
with respect to the stars in the field around the binary, the orbits of
the individual components may also be computed.

About 200 preliminary orbits have been determined for visual
binary stars, and of these about 100 are definitive. If, in addition to
the orbit, we know the parallax of the system, we can then compute
the size of the orbit.

It is not always necessary to see the faint secondary star, or

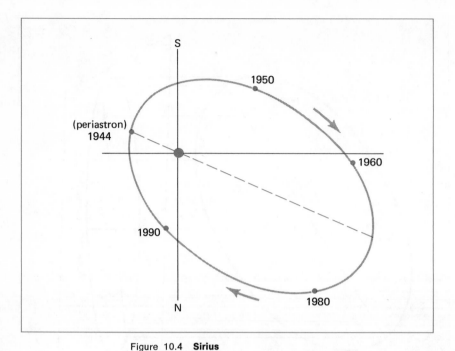

Figure 10.4 **Sirius**
Before Sirius B, the companion star of this system, was actually observed in the 1860s, astronomers knew that Sirius was a double star because of the bright primary's wavy proper motion. The companion is a white dwarf with a mass about equal to that of the sun. The two components of the system are now near their maximum separation as seen from the earth. Their period of revolution is 50 years. Recent precise measurements of Sirius by G. A. van Alhada and I. W. Lindenblad have revived the suspicion that there are small irregularities in the orbital motion, perhaps indicating a third body. Adapted from diagram by Jean Meeus in *Sky and Telescope*, vol. 41 (Jan. 1971):1, p. 23.

companion, in order to realize that one is observing a binary star. Even if the companion is too faint to be noticeable, the brighter component may still have a sinusoidal motion in the sky. The fact that its proper motion is not in a straight line, but is wavy, reveals that it is accompanied by a fairly massive but invisible companion. A number of these systems, called *astrometric binaries*, have been observed. Peter van de Kamp has done extensive work in this area and has discovered a number of such stars. In some cases, the wavy motion of a component of a binary system has led to the assumption that it is part of such a system and this assumption has later been confirmed by actual observation of the faint secondary.

In some cases, what appears to be a double star turns out to be a multiple star system. Castor is one example. Viewed through a small telescope, it separates into two stars, each of which on examination with larger equipment turns out to be a pair of stars. In addition, there is a remote companion revolving about the double pair in a large orbit, and this companion is itself double, thus making Castor a system of six stars. Epsilon Lyrae is another example. Viewed through a small telescope, it appears to be a double star. Viewed through a large telescope, each of its components turns out to be itself double.

Another binary star of particular interest is Sirius (Figure 10.4), the brightest star in the heavens. As far back as 1844 Friedrich Bessel had noticed that its proper motion was wavy and suspected it to be a double star with an invisible companion. In 1862, Alvin Clark, an American telescope-maker testing an 18½-inch lens, discovered a small companion to Sirius, a type of star known as a *white dwarf*.

## 10.2 SPECTROSCOPIC BINARY STARS

Not all binary systems can be resolved visually into two distinct stars. Many are observed by means of the spectroscope, using the principle of the Doppler shift. If the orbital plane of a binary system is not perpendicular to the line of sight, the components alternately approach and recede from the earth. Whatever the space velocity of the system's center of mass, the lines of the spectrum oscillate because of this orbital motion. Both spectra can be observed only if the stars do not differ in brightness by more than 1 magnitude. If the difference between the two stars is more than 1 magnitude, only the oscillatory line of the brighter component is observable. Figure 10.5 shows the spectrum of Mizar, a spectroscopic binary with a period of 20.5 days.

If the velocity of approach or recession is plotted against time, the result is a velocity curve. If both spectra can be observed, two

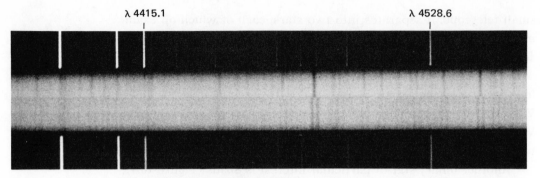

λ 4415.1                                    λ 4528.6

Figure 10.5 **The Spectrum of Mizar**
The upper spectrum was made on June 11, 1927. The lines are single, since
the components of the binary were moving across the line of sight. In the lower
spectrum, made on June 13, the lines are double, since one component is moving
toward, and the other away from, the observer. The difference in velocity due to
the orbital velocity of the components is 87 miles per second. Photographs from
Hale Observatories.

curves (Figure 10.6) can be plotted. They will be out of phase by
half a period and, depending on the masses of the two stars, will
generally have a different amplitude. The velocity curve gives us the
projected orbit of the star, not the true orbit. Since the projected
orbit is inclined to the plane of the sky, the true orbit cannot be obtained unless the inclination is known. This is available only if the
star also happens to be a visual or an eclipsing binary. The individual masses of the components cannot be obtained from the spectroscopic orbit alone, although lower limits for the masses may be.

Generally, the periods of revolution of spectroscopic binaries
are shorter than those of visual binaries. Visual binaries have been
detected with periods of perhaps over 10,000 years. Periods of spectroscopic binaries are generally less than 5 years, and over half of
them have periods of less than 10 days. This is because, other things
being equal, the shorter the period, the more rapid the movement of
the stars and the larger the Doppler shift. Hence it is easier to observe such systems spectroscopically. In some spectroscopic binaries, the components are so close together that gas streams from
the stellar atmospheres make confusion in measuring the shift of

294

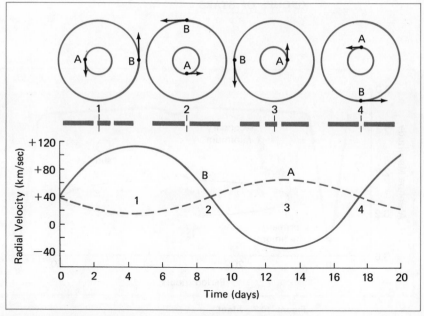

Figure 10.6 **A Hypothetical Spectroscopic Binary**
This example of a spectroscopic binary assumes that star A is more massive than
star B and that their orbits are circular. Velocity curves are plotted for one entire
revolution. The broken curve is that of the more massive star, A, and the solid curve
is for that of the star B. When the stars are moving perpendicular to the line of
sight (positions 2 and 4), their radial velocity is simply the speed of the entire system.
When the stars are coming toward or moving away from the observer (positions 1
and 3), their orbital velocities are superimposed on the radial velocity of the binary
as a whole.

the lines in the spectrum inevitable and an accurate orbit cannot be
determined.

## 10.3 ECLIPSING BINARY STARS

If the orientation of the orbit of a close pair of stars is nearly or
exactly on the line of sight or perpendicular to the plane of the sky
(which amounts to the same thing), one star of the system will
periodically eclipse the other. For obvious reasons, this phenomenon
is known as an *eclipsing binary*. The first eclipsing binary to be recognized was Algol (Figure 10.7), in the eighteenth century. The word
*Algol* means "demon," and ancient observers imagined that this star
was the eye of a demon winking in the sky. Its period of revolution

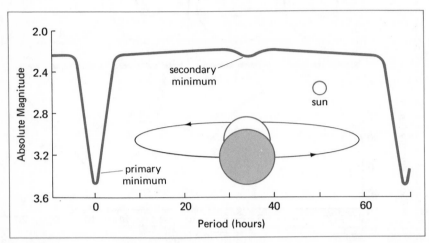

Figure 10.7 **Algol**
The diagram shows the light curve and the orbit of the eclipsing binary Algol. The sun is drawn to the same scale for comparison. Note that the darker star is larger than the brighter one. The eclipse of the bright component produces the greatest loss of light, or the primary minimum of the curve. The same area of the darker star is covered during an eclipse, but the loss of light is less, producing a shallow secondary minimum. (Light curve and orbit by Joel Stebbins.)

is 2 days and 21 hours. In 1889, spectroscopic observations of Algol confirmed that it was an eclipsing system of two stars. The orbit is almost on edge as seen from the earth, being inclined about 8°. Both stars are larger than the sun. The brighter one has a diameter three times that of the sun, and the dimmer star is still larger, by some 20 percent. The larger star is, however, only about 8 percent as bright as the smaller one. They are accompanied by a more distant small companion that revolves around the center of mass of the system in a little under two years.

Since the system is oriented so that the orbit is seen nearly on edge between eclipses, the Doppler shift in the spectral lines of the bright star reaches a maximum and a spectroscopic orbit can be obtained. After the light curve has been plotted, it is possible, from a combination of photometric measurements and spectroscopic observations, to obtain a great deal of information about the system and about the individual stars.

296

The periods of revolution of eclipsing binary stars range from 27 years to as little as 80 minutes. A number of phenomena affect the shape of the light curve; and, consequently, a detailed examination of the shape of these curves can reveal a great many things. If, for example, the orbit is circular, the space between the two minima will be equal. Also, the same area of each star will be eclipsed by the other. If the stars have different temperatures, an eclipse of the hotter star will produce a greater drop in brightness than the eclipse of an equal area of the cooler star. The deeper trough in the resultant light curve is called the *primary minimum*, and the shallower one, the *secondary minimum*. If the eclipse is total for an appreciable length of time, the minima will be flat on the bottom unless there is darkening at the limb, in which case the eclipse of the limb will result in less loss of light than the eclipse of the central portion. In this case, the curve will be rounded in the trough instead of flat. The amount of time from the first to the second contact, appropriately adjusted, gives us the diameter of the eclipsing star in terms of the dimensions of the orbit. When spectroscopic velocity curves are available, even more information can be obtained, including the inclination of the orbit to the plane of the sky (which is always near 90°), the size of the stars in terms of the sun's diameter, and the masses of the components in terms of the sun's mass.

If the stars are close enough together, other effects are observable. The close hemispheres may heat each other, and a peak may be observed in the light curve just before the beginning of the primary minimum. If tidal action has distorted the stars so that instead of being spherical they are *prolate spheroids* (something like footballs) with their long axes in line, the stars present a greater area to an observer on the earth midway between the primary and secondary minima. The top of the curve is then not flat, but convex upward.

We have assumed, in our analysis, that the orbits of binaries are circular, but frequently this is not true. If the orbit is definitely el-

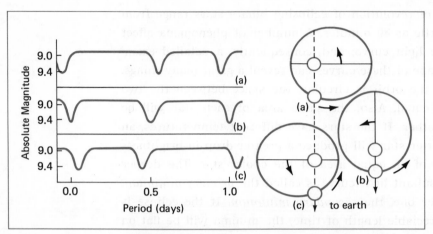

**Figure 10.8** **Light Curves of an Eclipsing Binary**
When the orbit of a binary as seen from the earth is nearly on edge, an eclipse occurs when one star passes between the observer and the other star. Because revolution in an ellipse is not uniform, the spacing of the eclipse minima in the light curve depends on the orientation of the major axis of the ellipse. A surprising amount of information regarding the components and their motion may be obtained from a study of the form of the light curve.

liptical, two effects are possible. If the eclipses occur at or near *periastron* and *apastron*, the points of smallest and largest separation between the stars, the eclipses will be of unequal duration because the stars will be moving most rapidly when they are closest together and most slowly when they are farthest apart. If the eclipses occur between these two points (and periastron and apastron occur between the eclipses), the intervals between the eclipses will not be equal because the stars will be moving more rapidly than usual during one interval and more slowly than usual during the other.

Algol has been used as an example of an eclipsing binary star. With a period of 2 days and 21 hours, stars three or four times the diameter of the sun eclipse one another. In many eclipsing binaries (as in Algol), the bright star is a Class B hot star and the fainter star is in a later spectral class but has a larger diameter. Curves of other types of eclipsing binaries are given in Figures 10.8 and 10.9.

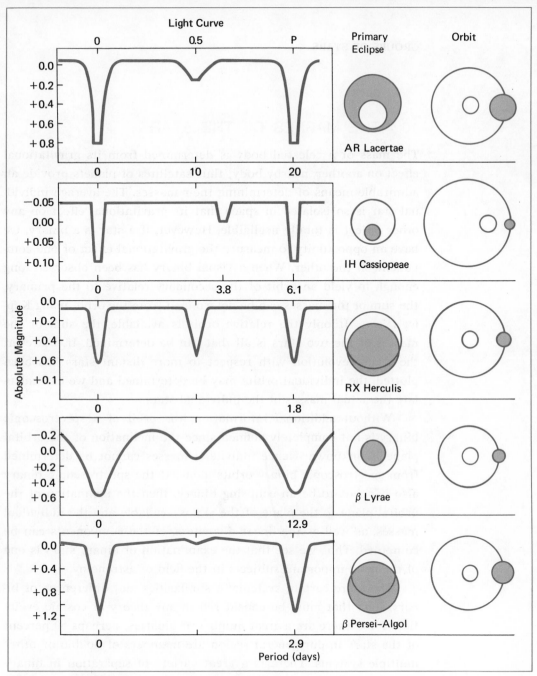

Figure 10.9 **Types of Light Curves and Associated Orbits**
When the eclipse is central, that is, when one star is either completely superimposed upon or occulted by the other star, the bottom of the minimum is flat. If one star is brighter than the other, one minimum is deeper than the other. In addition, the effect of tidal distortion (the elongation of the stars along the lines joining their centers) gives a characteristic shape to the light curve.

## 10.4 THE MASSES OF THE STARS

The mass of a celestial body is determined from its gravitational effect on another, nearby body; thus satellites of planets provide an admirable means of determining their masses. The average individual star is so isolated in space that its gravitational effect on any other object is totally negligible. However, if a star is a binary, we have an opportunity to measure the gravitational effect of one component on the other. When a visual binary has been observed long enough to yield an orbit of the secondary relative to the primary, the sum of the masses can be determined from the orbit, using Kepler's laws. If only the relative orbit is available, the sum of the masses of the two stars is all that can be determined. If, however, the star's revolution with respect to more distant stars has been plotted, the individual orbits may be determined and we can calculate the actual masses of the individual stars.

Without additional information, the orbit of a spectroscopic binary is not completely defined since the inclination of the orbital plane is unknown. Hence individual masses cannot be determined from spectroscopic binary orbits alone. If the spectroscopic binary also happens to be an eclipsing binary, then the inclination of the orbital plane to the plane of the sky is available and the individual masses, as well as the linear diameters, of the components can be computed. Thus we see that the examination of binary stars is one of the most important subjects in the field of astronomy.

There are certain systematic similarities and differences of binary stars that must be considered in any theory of cosmic evolution. First, there are a great number of binaries; perhaps 50 percent of the stars in our general region are members of double or other multiple systems. There is a great variety of separation in binary systems. Those with components almost in contact revolve quite rapidly; those with components separated by great distances move so slowly that their orbital motions have not been observed. As the separation of the components increases, so does the eccentricity of

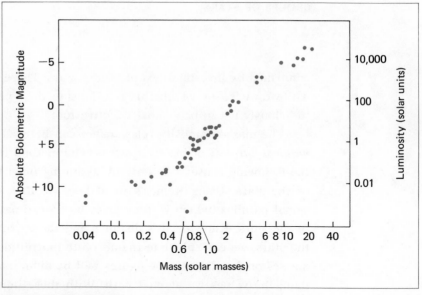

Figure 10.10  **The Mass-Luminosity Relationship**
When the masses of a large number of individual stars have been determined and their absolute magnitudes have been calculated from their known distances and apparent magnitudes, a correlation is found to exist, in most cases, between the mass and the intrinsic luminosity. There is a vertical scatter from the curve, and white dwarfs are well below the curve. For most of the stars studied, however, the curve (which is empirically determined) gives a fair approximation of the mass of the star when its absolute luminosity is known. (*Bolometric magnitude* is a measure of all the electromagnetic radiation emitted by a star.)

the star's elliptical orbit. Very close binaries have been observed to have extended gaseous atmospheres. Finally, there is a correlation, long known, between the differences in spectral type and brightness. If the two stars are main-sequence stars, then the brighter star is the hotter; if they are giants, the cooler (redder) star is the brighter of the two.

## 10.5 THE MASS-LUMINOSITY RELATIONSHIP

Since binary stars reveal the masses of their components in terms of the sun's mass and the parallax (and consequently the absolute magnitude) of the components is also available, these two variables can be plotted as in Figure 10.10. Thus a curve can be drawn relating the mass and the luminosity of a considerable number of stars. This *mass-luminosity relationship* was predicted theoretically and

301

confirmed by investigations of binary stars. There are stars, notably white dwarfs, for which this relationship does not hold; but used judiciously it can be a most effective tool.

The mass-luminosity relationship can be used to obtain the *dynamical parallax* of binary systems. The method used is iterative in the following sense: We start by assuming or guessing the masses of the stars. Using Kepler's third law, we then compute a provisional parallax, which in turn gives us an estimate of the distance to the stars. Knowing this distance and the apparent magnitude of the stars, we can calculate the absolute magnitude and look up the masses of the stars. These figures will be different from, and better than, the values we started with. With these improved values, we can return to the first step and repeat the entire procedure until the process stabilizes and gives us the individual masses of the components. The corresponding parallax, in seconds of arc, which is derived in obtaining these masses, is called the dynamical parallax of the system. This parallax has an error of approximately 5 percent. It can be obtained, of course, only for binary systems that are close enough to give us the orbital information needed to calculate it.

The masses of the stars do not differ widely from the mass of the sun. Indeed, the stars probably differ less in mass than in any other physical characteristic. Theoretically, a star with a mass greater than 65 times that of the sun could not exist, and a mass of much less than 0.01 the mass of the sun would be insufficient for the steady production of the stellar energy on any appreciable scale. This range of mass is effectively much smaller than the ranges of stellar temperatures, densities, and linear diameters that we have discussed.

## 10.6 GALACTIC CLUSTERS

We have considered multiple systems of stars with as many as six members. This does not exhaust the tendency of stars to group to-

Figure 10.11 **The Pleiades**
The Pleiades are a galactic cluster in Taurus. The stars are surrounded by gas and dust, which reflect the starlight. Photograph from the Hale Observatories.

gether. An examination of the sky also reveals the existence of *clusters,* or groups of stars apparently close together. Examples are the Pleiades in Taurus (Figure 10.11) and the Hyades in the same constellation. Since these groups of stars are for the most part in or near the Milky Way, they are called *galactic clusters.* Because their population is small and the distance between the stars rather generous, they are sometimes called *open,* or *loose, clusters.* When examined, they appear to share a common proper motion; and in this context they are referred to as *moving clusters.* Presumably, too, all the stars of a cluster were formed at about the same time from the same large cloud of gas.

A little over 40 years ago, Robert Trumpler of the Lick Observatory came upon a seemingly strange feature of the distribution of clusters. According to the then-accepted distances and dimensions of the clusters, the farther a cluster was from the earth, the larger its real dimensions were. This strange situation placed the earth at the core of a symmetric cluster system whose existence was inexplicable. Trumpler therefore concluded that there was absorbing material in and near the plane of the Milky Way (that is, in the direction of the clusters he was studying). The luminosity parallaxes on which the distances depended had been derived with the assumption that space was empty. If absorbing material was present, all clusters would appear fainter than they otherwise would and the luminosity parallax would place the clusters farther away from the observer. Since the angular diameter and distance of the clusters determined their linear size, the erroneous figure for their distances would result in overestimates of their sizes. For example, an absorbing medium that dimmed the stars by 1 magnitude, or a factor of 2.5, would cause an overestimate of their distance of 60 percent. A dimming of 2 magnitudes would increase the figure for the distance by about 150 percent, and so on. If the position of these clusters was used to outline the galaxy or a part of it, the result would be an oversized galaxy. Trumpler's work enabled astronomers to reduce the size of the Milky Way at a time when it seemed embarrassingly large compared with other systems.

One way of determining the distance of a cluster is from a color-magnitude diagram. If the stars of a cluster are plotted on an H-R diagram using the stars' color index (which is more easily determined than their spectral class) and apparent magnitude (which can be obtained from observations without knowing the distance of

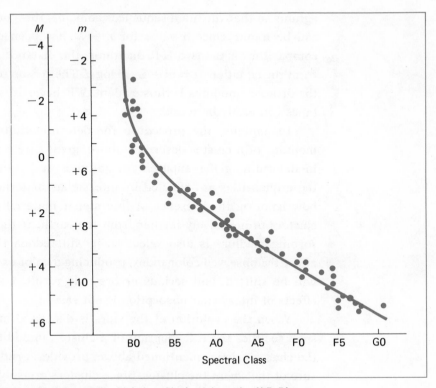

**Figure 10.12  The Distance Modulus of the Pleiades, Obtained from the H-R Diagram**
The points representing the stars of the cluster are plotted using the stars' observed
apparent visual magnitudes (*m*) and their spectral class. The line represents the
standard main sequence plotted from the stars' absolute visual magnitude (*M*) and
their spectral class. The modulus, $m - M$, = 5.5, indicating that the cluster is about
125 pc away.

the cluster), the stars generally form a main sequence from the
upper left to the lower right, although in some instances the sequence may not be complete. However, the lower portion will most
likely be intact. Since apparent rather than absolute magnitudes are
used, the distance still cannot be determined. However, if a standard
H-R diagram is matched to the diagram for a particular cluster (Figure 10.12), the magnitudes in the two diagrams will differ by a constant amount. This amount, which is equal to $m - M$, is called the
*distance modulus* of the cluster, and by positioning the two H-R di-

**305**

agrams so that the main sequences coincide, this constant difference can be found. Once a value for $m - M$ has been obtained from a comparison of the two H-R diagrams, the distance can be obtained from the equation $M = m + 5 - 5 \log d$. The reason for calling $m - M$ the distance modulus is thus evident. Whenever it is known, the distance can easily be found.

In applying the procedure for determination of the distance modulus of a cluster described above, great care must be exercised in determining the apparent magnitudes and color indexes. Both these quantities are affected by interstellar material. We have seen how ignoring the presence of interstellar material placed the open clusters progressively farther from the earth. If the interstellar absorbing medium is also selective, it will redden the starlight and affect the observed color index, producing a color excess. Both scales will be shifted, and serious errors may result. Correcting for the effects of interstellar absorption is not easy.

When the evolution of the stars is discussed in Chapter 11, we shall see that the H-R diagram of a cluster, in addition to revealing the characteristics mentioned above, provides a rather reliable measure of the age of the cluster. Since clusters are subject to the gravitational attraction of the rest of the galaxy, they cannot be considered permanent features. A compact cluster should last longer than a loose one, and a cluster far from the galactic plane should be more stable than one near the galactic plane where the tidal action would be more likely to disrupt its structure. Inevitably, however, individual stars in a loose or open cluster can be expected to escape from the cluster; and gradually, over millions or billions of years, the structure of the cluster will break down until it is completely dissipated.

Another clusterlike galactic aggregation of stars is the *association,* a loose aggregation of young stars with presumably a common origin and apparently a common proper motion. Although the stability of an association does not approach that of a galactic cluster, in its early stages an association looks and acts like a small, very open cluster. Presumably, it will dissipate in a fairly short time, astronomically speaking.

## 10.7 GLOBULAR CLUSTERS

The galactic clusters discussed in Section 10.6 are relatively small aggregations of stars within our own galaxy. All the known clusters of this type are within 3 kpc of the sun, since their low luminosity makes them invisible beyond this distance. Another type of cluster, with an entirely different distribution of stars and visible at distances greater than 4 kpc, is known as the *globular cluster*. As its name indicates, it is globular in form. It contains a very large number of stars, and, unlike many of the open clusters, it is free of any significant amount of interstellar gas and dust. Globular clusters look something like large chrysanthemums and, when viewed through a large telescope, are very beautiful.

Among the closest globular clusters are 47 Tucanae, at 4.6 kpc, and Omega Centauri (Figure 10.13) at 4.8 kpc. Better known in northern latitudes is M13, in the constellation of Hercules, at a distance of about 8 kpc (Figure 10.14). More than a hundred globular clusters have been discovered in our galaxy, and several hundred may exist.

The distances of nearby globular clusters were originally obtained from observations of RR Lyrae stars. When work was first done in this area, RR Lyrae stars were thought to have an absolute median magnitude of about zero. It now appears that their absolute median magnitude is probably nearer to 0.6. Any change in this assumption will change the distances of all the globular clusters. All the known clusters are farther than 3 kpc's away, too far for direct parallax measurements. In the more distant clusters, the RR Lyrae stars are too faint to register, even in the largest telescopes. Harlow Shapley, in his work on globular clusters, adopted the principle of examining the brightest stars in a cluster, which are about $1\frac{1}{2}$ mag-

Figure 10.13 **The Globular Cluster Omega Centauri**
This cluster is at a distance of about 4.8 kpc and is slightly oblate, presumably as a result of rotation. Harvard Observatory photograph.

nitudes brighter than the RR Lyrae stars. On the basis of the information thus obtained and assuming that statistically the globular clusters all had the same diameter, he estimated the distances of those that could not be reliably resolved into individual stars on the basis of their apparent angular diameter. According to these calculations, the most distant globular clusters are 50 or more kpc from us. Figure 10.15 shows the distribution of globular clusters with respect to the sun.

Globular clusters contain a very large number of stars. A hundred thousand, perhaps even a million, stars may be present in an average cluster. Investigations of the mass of globular clusters indi-

Figure 10.14 **The Globular Cluster M13**
This cluster is in Hercules and is about 8 kpc from the solar system. Photograph from the Hale Observatories.

cate that it is approximately 1 million times that of the sun. Of course, not all the stars can be seen. Hale Observatories astronomer Allan Sandage counted 44,000 stars in a photograph of the globular cluster M3 taken with the 200-inch telescope, but many more were uncountable in the close-packed nucleus. Considering that most globular clusters are between 10 and 30 pc in diameter (their maximum diameter being about 100 pc) and that they are highly concentrated toward the center, the average density of stars in the clusters must be about a thousand times the density of stars in the sun's neighborhood. In the center, there may be as many as 50,000 times this figure.

The great distance of globular clusters from the earth makes a study of their internal properties very difficult. Most are probably oblate. The proper motions of a few clusters have been measured, and estimates of their space velocity range to above 125 miles per second relative to the sun.

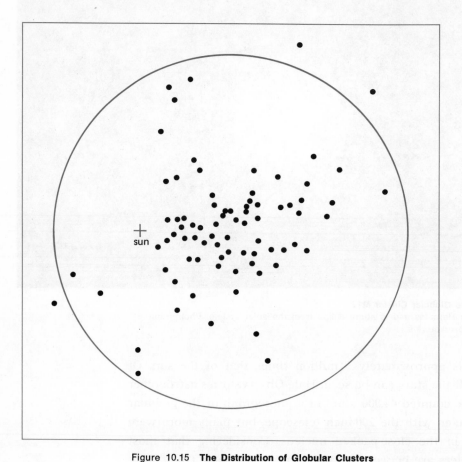

**Figure 10.15  The Distribution of Globular Clusters**
The circle represents the approximate outer edge of the Milky Way system. The
globular clusters are projected vertically onto the plane of the Milky Way, although
they may be far above or below it. The sun's approximate location in our spiral is
indicated by a cross. Note that most of the globular clusters are in the half of the sky
toward the center of the Milky Way as seen from the sun. It is this that accounts
for the apparent distribution of globular clusters in the terrestrial sky. The radius of
the circle is about 20 kpc. From "Globular Cluster Stars" by Icko Iben, Jr.
Copyright © 1970 by Scientific American, Inc. All rights reserved.

The average lifetime of a globular cluster should be rather long
compared with that of an open galactic cluster, in part because of
the high concentration of matter near the center, which lends considerable stability to the structure and prevents the escape of stars

near the periphery, and in part because the average globular cluster is outside the principal plane of gravitational action by the mass of the galaxy.

## 10.8 STELLAR POPULATIONS

The *galaxies*, which are the largest groups of stars, will be treated in later chapters. However, some information about our own Milky Way galaxy is needed here. A flat spiral structure containing about 100 billion stars, it has a concentrated *nucleus* with *spiral arms* in the principal plane, wrapped about the nucleus. The hemispherical regions above and below the principal plane make up what is called the *halo*, and the nucleus and arms are known as the *disk*. The accepted diameter of the disk is about 30 kpc. The sun is in one of the spiral arms, about 8 or 9 kpc from the center.

As early as 1915, Harlow Shapley noticed that the H-R diagrams for stars of globular clusters differed greatly from the conventional H-R diagram proposed by Russell. About 30 years later, Walter Baade proposed the concept of two populations of stars. Population I consisted of the stars whose H-R diagrams resembled Russell's early diagram. Population II was made up of stars in the globular clusters in the nuclei of extragalactic systems and in the elliptical galaxies. The brightest Population I stars were those at the upper end of the main sequence, the Class A, B, and O giants with an absolute magnitude of −4 or brighter. The brightest Population II stars were red giants. Since Population I stars were found in regions with abundant gas and dust and Population II stars in areas devoid of gas and dust, astronomers concluded that star formation did not occur in Population II regions and hence that this population was old compared with Population I. Stars in the sun's neighborhood appeared to belong to Population I. Baade's original division of stars into two populations was perhaps too simple, but it led to further discussion of the differences in character and prop-

**311**

erties of different groups of stars. In 1957, as the result of a conference held at the Vatican Observatory, five, rather than two, population groups were suggested. These five populations are summarized by A. H. Oort as follows:

1. Halo Population II. This group contains the oldest stars. It is represented by the oldest globular clusters and separate stars in the galactic halo. An age as great as 10 billion years may be assigned to this group.
2. Intermediate Population II. These stars are found in the halo and particularly in the nucleus of our galaxy.
3. Disk Population. These stars are found in the arms and in the central region of the galaxy. The sun is a member of this population. The age of this group may be as great as 5 billion years.
4. Intermediate Population I. These are massive, hot stars younger than the sun, such as Sirius. They are found in the principal plane of the galaxy.
5. Extreme Population I. This is a young population, found in the spiral arms of the galaxy and in regions of considerable gas and dust. These stars are typically massive and hot, usually Classes O, B, or A giants.

As we progress through the five classes, starting with the first, we find a tendency toward concentration nearer the plane of the galaxy. Halo Population II is found in the regions above and below the plane, averaging as much as 2 kpc from the galactic plane. Extreme Population I tends to be near the plane. The first two classes are concentrated toward the center, whereas the later ones show no such preference and are found largly in the spiral arms. Another distribution is found when we examine the ratio of heavier elements to hydrogen. As we progress through the classes, this proportion seems to increase. Otto Struve has suggested that this implies the possibility that more recently formed stars were produced from interstellar material richer in heavy elements produced by older and now defunct stars.

Our knowledge of stellar populations is still far from complete, and further research in this area is expected.

# QUESTIONS

(1) What is a visual binary? (Section 10.1)

(2) What is the difference between the apparent relative orbit and the true relative orbit of a visual binary? (Section 10.1)

(3) What is an astrometric binary? (Section 10.1)

(4) How do astronomers detect spectroscopic binaries? (Section 10.2)

(5) What are the primary and secondary minima in the light curve of an eclipsing binary? (Section 10.3)

(6) What can we learn about the stars in an eclipsing binary from examining the shape of its light curve? (Section 10.3)

(7) What is the mass-luminosity relationship, and why is it important? (Section 10.5)

(8) Why was the existence of absorbing material in the interstellar medium inferred from observations of galactic clusters? (Section 10.6)

(9) Describe a typical globular cluster. (Section 10.7)

(10) How are Population I and Population II stars different? (Section 10.8)

# CHAPTER 11

# GAS AND DUST: STELLAR EVOLUTION

Our sun is located in one of the spiral arms of our galaxy, a little over halfway from the nucleus to the edge. About 50 percent of the mass of the material in this region is in the form of gas and dust; the other 50 percent is in the stars. By contrast, the halo above and below the spiral seems to be relatively free of any material except stars. The average cloud, or nebula, in our neighborhood is probably about 99 percent gas and 1 percent dust, although an exact evaluation of this proportion is impossible. The two materials react quite differently. Gas absorbs and reemits energy and in appropriate circumstances it will produce either bright- or dark-line spectra. Dust produces no lines in the spectrum, but scatters, reddens, and dims the light. In this chapter we shall consider the various types of nebulas found in our galaxy and their reactions to a variety of conditions.

Since most modern theories of the evolution of the sun are based on the behavior of nebular material, we shall also consider, in this chapter, the subject of stellar evolution. We shall first trace what appears to be the probable life history of a star similar to the sun and then consider the evolution of stars of different characteristics. In addition, we shall study the end-products of ordinary stars with different masses, which have a wide range of characteristics.

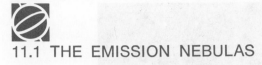

## 11.1 THE EMISSION NEBULAS

The word *nebula* literally means a cloud of gas. At one time, any hazy patch in the sky was referred to as a nebula. Messier's catalog of over a hundred nebulas includes every type in the former sense of the term. A careful examination of these fuzzy patches with large telescopes shows that they differ widely. In addition to clusters of stars, there are also *galaxies* ("Milky Ways" outside of our own) and true nebulas. A true nebula is actually a cloud of gas.

The best-known and brightest of the emission, or bright-line, nebulas is in the constellation of Orion (Figure 11.1). Known as Theta Orionis, it appears to be the middle star of three in the sword of the giant. It appears slightly fuzzy and quite faint to the naked eye, but can be examined in considerable detail with a large telescope. It looks green to the eye, because of strong emissions in the green part of the visible spectrum, and resembles a turbulent cloud frozen in action. The lack of apparent movement is due, of course, to its distance.

The Orion nebula is 500 pc from the earth. The central portion, although its boundary has not been precisely defined, is about 3 pc in diameter, and the overall diameter is about 10 pc. The very brightest portion, which is all that can be seen with a small telescope, is considerably smaller. Long-exposure photographs taken with large telescopes are necessary to reveal the fainter portions. The light emitted by the nebula comes from the radiation of a number of Class O stars. Four stars particularly noticeable in a small telescope or a short-exposure photograph are called the *Trapezium*. Longer-exposure photographs show more stars.

The source of light for all bright-line nebulas is one or more Class B1 or hotter stars. These nebulas have, as their name indicates, a bright-line spectrum. The strongest emissions come from certain "forbidden" lines of ionized oxygen, nitrogen, and neon. In the early part of the twentieth century, the chemical source of these lines was a considerable mystery. However, in 1927 Ira Bowen explained the forbidden lines as the result of transitions in common atoms that rarely or never occur under terrestrial laboratory con-

Figure 11.1 **The Great Nebula in Orion**
This photograph was taken with a 100-inch telescope. Most of the hot stars which supply the energy that makes this emission nebula visible are obscured by the glowing gases. Photograph from the Hale Observatories.

ditions. The nature of the stellar light source has a considerable effect on the radiation received from the nebula. The brightness of a nebula, for example, is a function of the brightness of the star or stars it contains. If there were no stars in a nebula, it would be completely dark. Because emission nebulas are found only near very hot stars, hydrogen is abundant in their spectra.

The source of energy in an emission nebula is, as we have said, a Class O or an early Class B star. These are among the hottest stars, and their maximum radiation is in the ultraviolet. Such radiation can easily ionize atoms. For example, radiation with a wavelength of less than 912 Å can remove the electrons from the hydro-gen atoms in a cloud of tenuous gas. After being freed from their parent atoms, these electrons are captured by other atoms deficient in electrons. They need not return to the energy state from which they were released, but can cascade downward through energy levels and reemit energy in the visible range. Since it is emissions from these atoms that produce radiation, the result is a bright line. This is why nebulas surrounding hot stars have bright-line spectra.

The physical conditions in any nebula are impossible to dupli-cate in a terrestrial laboratory. Nebulas are, on the average, very tenuous, being a better vacuum than that which can be produced in a laboratory. They also have great depth, and although few atoms are to be found per cubic mile, many parsecs of nebular material may lie in the line of sight, thus allowing us to see the phenomena that occurs. The Orion nebula, for example, has a very low density but a large diameter. Intrinsically, it is faint. Interestingly enough, if we traveled through space until we were close to the Orion nebula, it would appear bigger, but no brighter per square degree than it does from the earth. Changing our distance from the nebula would change the total amount of light we received from it in accordance with the inverse-square law. But although the total amount of light would increase, the apparent area would increase by exactly the same amount and the brightness per unit of area would remain the same.

## 11.2 REFLECTION NEBULAS

If the star in a nebula is cooler than about a Class B1 star, there will be no emission lines in its spectrum and all the illumination will be in the form of reflected starlight. Of course, even in an emis-sion nebula, there is some reflection, since the physical composition of the two types of nebulas is about the same. However, the bright lines of the emission spectrum overcome the lower level of illumina-

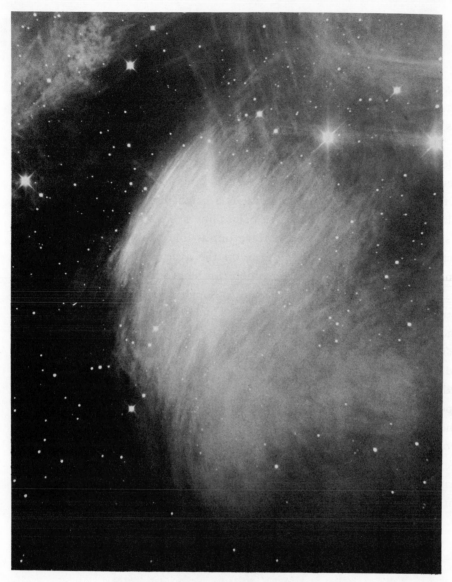

Figure 11.2 **A Nebulosity in the Pleiades**
This photograph, taken with a 60-inch telescope, shows a reflection nebula around
Merope, one of the Pleiades. Photograph from the Hale Observatories.

tion of the continuous background. Reflection occurs because of the
presence of dust particles. Although dust makes up a minor part of
the mass of the nebula, the particles are essential to the reflection
of starlight. What material these particles are made of is unknown,

318

since, unlike the gas that produces the emission lines, the dust particles do nothing but reflect the light of the star. This light thus has the usual dark-line spectrum, identical or nearly so to the spectrum of the star itself. It has been suggested that the reflective power of the solid particles in a nebula would be enhanced if they were water crystals or had a frost or ice cover.

The nebulosity surrounding the stars of the Pleiades (Figure 11.2), notably Alcyone, is an example of a reflection nebula. The Pleiades stars are relatively cool and lack sufficient ultraviolet radiation to ionize the gases of the nebula significantly. Another example of a reflection nebula is that surrounding the bright star Antares. This is an M0 star of luminosity Class I, a supergiant. The nebula is clearly visible when photographed in red light but appears insignificant in blue light. The same is true, to some extent, of the star itself, since it is very red.

Although we cannot be sure the internal material of a reflection nebula closely resembles that of an emission nebula, it is reasonable to assume that it does. The difference is in the star and in the reaction of the nebular material to the differences in maximum radiation of the stellar spectra. In all but a very few nebulas, either the emission or the reflection type, one or more identifiable stars is responsible for the energy that makes the cloud visible.

## 11.3 DARK NEBULAS

Throughout the sky, particularly in the Milky Way, there are many examples of what were at one time thought to be "holes in the heavens." Near the Southern Cross, for example, is an area so dark that it has been named the Coalsack. Dark lanes in the Milky Way are plainly evident on a dark night, particularly in the general region of the constellation of Cygnus. When the Milky Way was first examined with telescopes, it was thought that a lack of detectable stars, particularly faint ones, indicated that stars were less numerous in a particular region than they were on the average in the Milky Way. Early attempts to construct a model of the galaxy were frustrated by this assumption. Indeed, when Herschel made his first tentative model of the Milky Way, he put a split in one side of it, omitting stars because he found none when he looked along the center line in that particular direction. It was not until the present century that the nature of the dark lanes and other apparent holes was understood. If we count the stars up to a certain magnitude and then count those 1 magnitude fainter, we find that the number of stars observed is not as great as it should be if they were uniformly distributed through space. Presumably, fainter stars are farther away; statistically, therefore, an increase of 1 magnitude in the range of our search should produce four times as many stars. This phenomenon is not observed in practice because the more distant stars (those of larger magnitude) shine through more absorbing material. Dark nebulas, then, are detected not by what they show, but by what they obscure.

Presumably, dark nebulas have the same physical structure and constitution as bright nebulas, but have no star or stars to illuminate them. A good illustration of the absorbing power of these clouds is provided by the *zone of avoidance*. As we shall see when we consider galaxies other than our own, they do not appear to exist in any significant numbers near the plane of the Milky Way, not even in the direction away from the galactic center, which is the shortest way out. The reason for the zone of avoidance is not an absence of galaxies in that direction, but the inability of electromagnetic radiation to penetrate the absorbing clouds in the Milky Way. Most of the known dark clouds are within 100 to 500 pc of the earth, and none is known with an established distance greater than 2 kpc. Of course, it is assumed that dark material exists throughout the spiral arms of the galaxy. The central region is completely obscured by this material. If the Milky Way were completely unobscured, it would appear many times brighter than it does, particularly in the summertime. Figures 11.3, 11.4, and 11.5 show various dark nebulas.

Figure 11.3 **A Nebula in Monoceros**
This photograph was taken in red light with a 48-inch Schmidt telescope. Note the many dark nebulas scattered throughout the area. Photograph from the Hale Observatories.

## 11.4 THE INTERSTELLAR MEDIUM

Even a casual glance at the Milky Way reveals a very irregular form containing dark areas where the number of stars is much lower than average. These dark lanes, as we mentioned in the preceding section, are now known to consist of interstellar material that cuts

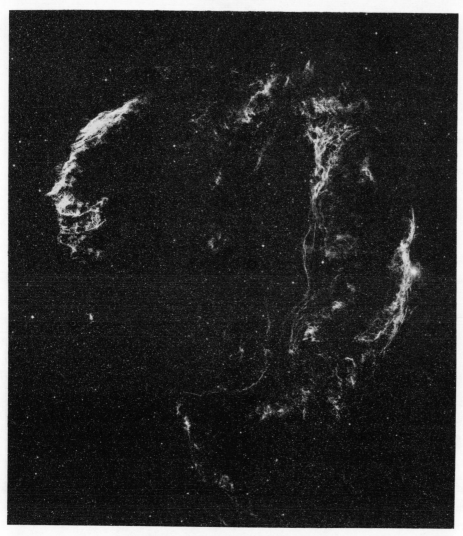

**Figure 11.4 A Filamentary Nebula in Cygnus**
Like Figure 11.3, this photograph was taken in red light with a 48-inch Schmidt telescope. This extensive nebula is evidently all from one source, perhaps a supernova. Photograph from the Hale Observatories.

off the light of the stars beyond. If the stars were distributed uniformly through space, simple star counts would enable us to measure the extent and density of this interstellar material. However, since the stars are nonuniformly distributed (decreasing in number, particularly, with increases in galactic latitude), simple star counts are not enough. Determining the amount of dimming caused by

322

Figure 11.5 **The Horsehead Nebula in Orion**
This photograph was taken in red light with a 200-inch telescope. The nebula is
south of Zeta Orionis. There is strong absorption at the bottom and some bright
material near the top. Photograph from the Hale Observatories.

darkening material and the actual diminution in the number of stars
owing to their nonuniform distribution is a complex problem. However, because of the phenomenon of selective absorption (since the
dust reddens the stars), we do have a clue to the amount of material
through which the starlight from a particular source passes.

In Chapter 10 we mentioned Robert Trumpler's 1930 work on
galactic clusters. Trumpler found a systematic arrangement of progressively larger clusters extending outward in all directions from
the earth. Since such an arrangement was untenable, he concluded
that something was dimming the light of the clusters and making
them appear too far away. He proposed the existence of absorbing
material and arrived at a provisional figure of 0.32 magnitude per
kpc for general absorption in the galactic plane. Among other

things, Trumpler found that stars with a small proper motion (and hence presumably distant) were systematically weak in ultraviolet radiation. Stars at a considerable distance from the earth would presumably shine through more of the interstellar medium, and selective absorption would diminish the blue, violet, and ultraviolet radiation from them.

Selective absorption produces a color excess, which we have defined as a numerical measure of the reddening due to absorption. The color index of the stars may be obtained by photoelectric or photographic photometry. If the photographic magnitude of a star is compared with its visual magnitude, the difference is the color index. If a spectrogram is made and the spectral class of the star is determined from the line pattern, a color index is automatically associated with that spectral class. If these two indexes are the same, there is no color excess; if they are not, the difference is the color excess of the star. The formula relating absolute and apparent magnitudes with distance may be expanded to

$$M = m + 5 - 5 \log d - K$$

where $K$ is the color excess. If $K$ is zero, there is no absorption. Color excess is extremely important in determining the distance of objects in the galactic plane, where absorption is greatest.

Various figures have been derived for the value of $K$. It must be carefully determined in each case, and no figure or function that will generally work can be given. Its determination is extremely important because a loss of 1 magnitude in the brightness of a star owing to absorption will mean an overestimate of its distance by about 60 percent if a luminosity parallax is determined without taking absorption into account. This is because a diminution of 1 magnitude means that the apparent brightness of the star is diminished by a factor of 2.5. Since the square root of 2.5 is approximately 1.6, the figure for the distance will be 1.6 times the true figure, or 60 percent too large. In certain areas of the sky, absorption may produce a diminution in brightness of as much as 6 magnitudes.

Absorption is greatest in the galactic plane. Shapley, in his study of the globular clusters, found that they underwent practically no reddening because they were generally seen above or below the Milky Way. Consequently, they were relatively unaffected by selective absorption, and, presumably, determinations of their distances from luminosity parallaxes would need little or no correction.

The nature of the material making up the obscuring clouds of interstellar medium is partially revealed by the spectroscope. Additional information can be obtained by examining the absorbing properties of the material. Atomic material produces dark lines in a spectrum, and molecules produce complex lines or bands. This process is, in a sense, selective absorption; but what is generally called selective absorption results from the dust particles in the cloud. If these particles are no larger than a hundred-thousandth of an inch in diameter, blue and violet light as well as ultraviolet is systematically subtracted from the electromagnetic radiation that passes through the cloud. Particles with larger diameters produce no selective absorption. If they are considerably larger, their general absorption is rather insignificant since their cross-sectional area (which determines their absorbing ability) is small in relation to their mass. The interstellar medium is estimated to be mostly, perhaps as much as 99 percent, gas. The dust particles are a very small proportion of the total material, but are most effective in reddening light.

By far the greatest part of the material in the universe—including the interstellar medium—is hydrogen. Helium and the heavier elements are only a small proportion of the total material. Clouds of hydrogen around bright Class O and B stars are excited and ionized out to a particular spherical limit. Consequently, they appear bright. If O and B giants are close enough together, more or less continuous regions of glowing gas are observed. These are called *H II regions* and are used for tracing both the arms of our own galaxy and the arms of exterior galaxies. An estimated 5 percent of the gas in the spiral arms of the galaxy is found in the H II regions.

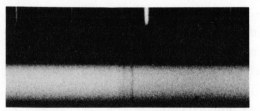

H line of calcium II

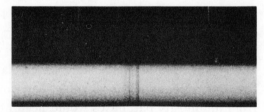

K line of calcium II

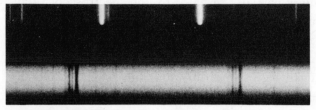

D lines of sodium I

**Figure 11.6  Interstellar Lines**
These calcium II and sodium I lines are superposed on the spectrum of Epsilon Orionis and split into five components. They indicate absorbing clouds between us and the star with velocities relative to the sun of +2.4, +7.0, +10.9, +15.4, and +17.1 miles per second. Spectra from the Hale Observatories.

Outside the limits of excitation by hot stars, the hydrogen in the interstellar material is optically invisible because it produces no emission lines and no significant absorption in the optical region. However, at a wavelength of 21 centimeters (in the radio range), there is an emission line due to neutral hydrogen. This line has proved invaluable in mapping the gas clouds, and hence the spiral arms, of our galaxy. It is discussed at length in Chapter 12.

In 1904, J. Hartman discovered a stationary line due to calcium II in the spectrum of the spectroscopic binary Delta Orionis that did not oscillate with the lines of the binary star. This line was also quite sharp. The suggestion that the line was due to a cloud of gas between Delta Orionis and the earth was not universally accepted at first, but it is now generally believed that such lines do represent clouds of absorbing gas. They are found mostly in or near the plane of the galaxy and are produced by, among other things, neutral sodium, potassium, calcium, iron, ionized titanium and calcium, and radicals of hydrogen, carbon, and nitrogen. Figure 11.6 shows calcium II and sodium I lines superimposed on the spectrum of Epsilon Orionis. Because of the different velocities associated with clouds between the earth and other light sources, multiple lines are also found. Turbulence within individual clouds may also produce a multiplicity or broadening of interstellar lines.

In many photographs of the Milky Way, there are small, dark specks that, upon examination, turn out to be globules of dark material. They take on different forms but are generally small in diameter, at most about 100,000 AU's, and may be stars in the process of formation.

## 11.5 INTERSTELLAR CHEMISTRY

Although interstellar atomic material has been studied intensively for decades, the molecules in the interstellar medium have been reluctant to reveal their secrets. The first molecules detected (the radicals CH, $CH^+$, and CN) were found over 30 years ago, by examining absorption bands in the spectra of hot stars. Since the early 1940s, the emphasis in this area has shifted from optical to radio astronomy.

In 1963, the hydroxyl diatomic molecule OH containing the common isotope of oxygen was detected, as was OH containing the heavier isotope of oxygen. Interstellar ammonia was discovered in 1968 and water vapor in 1969. These discoveries were followed by the detection of formaldehyde ($H_2CO$), carbon monoxide (CO), cyanogen (CN), which was rediscovered by radio techniques, hydrogen cyanide (HCN), molecular hydrogen ($H_2$), and cyanoacetylene ($HC_3N$). All the molecules discovered since 1960 have been found by radio astronomers searching the microwave (millimeter-centimeter) region of the electromagnetic spectrum, and extensive research of this type can be expected in the future.

## 11.6 THE EVOLUTION OF THE STARS

Historically, all theories of the evolution of the stars have started with a contracting ball of gas. Indeed, during the nineteenth century, the physicist Hermann von Helmholtz suggested that contraction accounted for the heat and light of the sun. Since the true age of the sun (and the total amount of energy it has generated and will generate during the present stage of its life cycle) was unsuspected at that time, astronomers took this theory quite seriously. The nebular theory still seems the most valid explanation of the birth of the stars, even though it does not account for their continuing energy output (which we now know to be due to nuclear processes in stellar interiors).

Our present beliefs about the life and times of the stars are derived from observation, logic, and intuition, aided by complex computer models of astronomical processes. We cannot, of course, observe the life, maturation, and death of a single star, since stars live for billions of years. Spread out before us in the galaxy, however, are stars of all ages, from the very youngest, immature stars, evidently just formed from a cloud of gas, and perhaps still accompanied by a portion of it), to the aged white dwarfs, which probably represent the last visible stage in the evolution of stars like the sun.

The conventional H-R diagram is very useful in studying stellar evolution. When a star changes its position on the diagram, this does not represent an actual movement through space but a change in the star's characteristics, as time passes and the process of evolution proceeds. Horizontal motion reflects a change in the star's spectral class, which is produced by a change in its temperature. As the star moves to the left, it is heating up; when it moves to the right, it is cooling off. Vertical movement indicates a change in luminosity. A star near the top of the diagram is intrinsically very luminous, because it has a high temperature or a large diameter or both. A star near the bottom is small and intrinsically faint. With these facts in mind, we can proceed to a discussion of the current theory of stellar evolution.

A star begins as a relatively cool cloud of gas, perhaps many times its ultimate mass. With the passage of time, the ball of gas begins to contract as a result of its own gravitational attraction. Its rate of rotation, if it is rotating, speeds up to conserve angular momentum, and the temperature in the center of the cloud rises. These young "protostars" are still very cool, and very large, typically thousands of AU's in diameter. Because of their large size, they are very bright. Thus a newborn star makes its first appearance in the upper right-hand part of the H-R diagram.

Young stars are extremely unstable. The first person to give a complete theoretical analysis of this stage of stellar evolution was the Japanese astrophysicist Chushiro Hayashi. Hayashi found that a new star contracts rapidly with no appreciable change in the temperature of its atmosphere. As a result, its luminosity decreases, but its spectral class remains unchanged. Thus the initial path followed by a newborn star on the H-R diagram is downward in a vertical line, sometimes called a *Hayashi line*.

While a star is moving down a Hayashi line, the dominant physical process occurring is convection. Energy is carried from the center of the star to its surface by leisurely currents. Gradually, however, the transport of energy by radiation rather than convection becomes more important. As this happens, the vertical descent of the star on the H-R diagram is halted and its path veers abruptly to the left; the surface temperature of the star has begun to rise. Finally, after thermonuclear reactions deep in the star's center are well underway, the star becomes quite stable and gradually approaches a part of the main sequence called the *zero-age main sequence*. After the star moves horizontally toward this boundary, it drops slightly, to the *zero-age line*. Where the star's horizontal motion terminates depends primarily on its mass, although other considerations (such as its chemical composition) have some effect. Stars with a large mass become Class O or B giants far up the main sequence and to the left of the spectral scale. Stars with a mass equal to that of the sun become Class G stars with a magnitude of +5. Stars with a small mass end up as faint red stars low in luminosity and temperature.

In the early part of this century, the main sequence itself was regarded as an evolutionary path. Astronomers suggested that perhaps all stars began as red giants, moved horizontally across the top of the H-R diagram as they contracted and heated up, until they become O or B giants, and then slid down the main sequence to the area of the red dwarfs.

Since stars shine by destroying the matter of which they consist, a star's lifetime should be proportional in some way to its mass. More massive stars having more material for the production of energy should, superficially at least, have the longest lives. Actually, stellar lifetimes do depend on the mass of the star, but the relationship is not so simple. The massive stars, although they have more material to convert into energy, have the shortest lifespans. That is because their temperatures, and consequently the rates at which they use up material and radiate energy, are much greater than those of stars with smaller masses. For example, the most massive stars, 30 or 40 times the mass of the sun, probably have lifespans of millions of years. The sun probably has a total main-sequence lifetime of about 10 billion years. A small star with only 10 percent the sun's mass may live a hundred times as long.

As a main-sequence star ages, it moves slightly upward on the H-R diagram because its luminosity is increasing. During this time, it may have central temperature of tens of millions of degrees, depending on its mass. The proton-proton reaction and, in the case of the hotter stars, the carbon cycle convert hydrogen to helium. The sun converts about $4.3 \times 10^{12}$ grams of hydrogen per second into energy. Obviously, this process cannot go on forever. Eventually the hydrogen supply in the core of the star will be exhausted. When this occurs, its main-sequence lifetime ends. During its residence on the main sequence, the star will have slowly brightened, perhaps by as much as a magnitude. This may be one of the reasons the main sequence has an appreciable width. The time spent on the main sequence may be different for different generations of stars and consequently for stars with different chemical compositions, but fundamentally it appears to depend on the star's mass. Massive stars stay only a short time on the main sequence; stars with a low mass remain a long time.

When the hydrogen in the core of a star is exhausted, the core is no longer able to maintain its previous dimensions and collapses. The shell outside the core then begins to convert hydrogen into helium at a furious rate. The temperature increases, and the shell expands. How great the expansion may be and what the governing factors are still constitute important subjects for research. Eventually, the star becomes a red giant in the upper right portion of the H-R diagram. How large it becomes depends on a variety of factors, but red giants as large as the orbit of Mars do exist. A star that expands so spectacularly also diminishes in density spectacularly. The typical red giant, although it may have started out with a mass several times that of the sun, is so large that its average density is extremely low.

The point at which the core of the star collapses and the outside of the star expands is called the *turnoff point*. At this stage, the star proceeds upward on the H-R diagram as its luminosity increases and moves to the right as its outside expands and cools. After a long period, the star stabilizes on the red-giant branch and increases in luminosity for perhaps 250 to 300 million years. When it reaches its brightest on this branch, the helium in the core begins to fuse and form carbon by a thermonuclear process called *helium burning*. Increasing temperatures accelerate this process until the core explodes in a "helium flash." The star then gradually moves from the red-giant branch in the H-R diagram to the *horizontal branch*, where it achieves a luminosity about 80 to 150 times that of the sun. It moves horizontally on the diagram at this stage, passing back and forth through the *instability strip*. During this

331

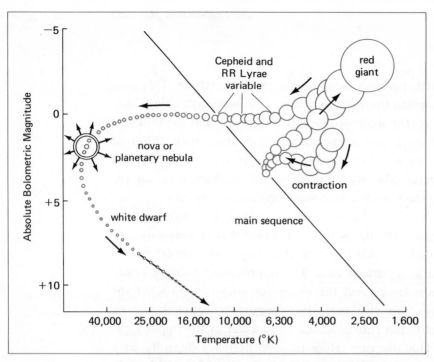

Figure 11.7 **The Evolution of a Star of the Sun's Mass**
This diagram shows the evolution of a star of 1 solar mass on the H-R diagram.
The size of the circle indicates, but is not accurately proportional to, the supposed
size of the developing star. Absolute bolometric magnitude is a measure of the
total electromagnetic radiation received from the star at a distance of 10 pc.

time, the star may become a Cepheid or an RR Lyrae variable. It
may make several trips across this branch. If it becomes sufficiently
unstable, its surface may be blown off to form a planetary nebula.
In any event, it eventually decreases in luminosity and drops down
to the left of and below the main sequence, at which point it is
classified as a *white dwarf,* or a *collapsed star.* Figure 11.7 shows
the evolution of a star of the sun's class by means of an H-R
diagram.

## 11.7 WHITE DWARFS

The first white dwarf found by astronomers was Sirius B, the
faint companion of the binary Dog Star, which was identified in

332

1862. Because it is part of a binary system, we can calculate the mass of Sirius B in terms of the sun's mass. We also know, from its low intrinsic luminosity (+11.5) and high temperature (about 12,000°K), that its diameter is very small.

Dwarf stars appear on the H-R diagram 9 or 10 magnitudes below the main sequence in the lower left corner. Most are white (hot), but a few are yellow and at least one is red. They are easily identified by their characteristic spectra, but their low luminosity (less than 1 percent the luminosity of the sun) makes them difficult to detect at large distances.

The most astonishing thing about the white dwarfs is their small diameter and consequent high density. Sirius B is about the mass of the sun but has a diameter more nearly that of the earth. This results in a density about a million times that of water. A tablespoon of matter from a white dwarf would weigh thousands of tons on the earth.

Since the diameter of a white dwarf is small and its mass large, the surface gravity must be enormous. One of the predictions of the theory of relativity is that electromagnetic radiation leaving a high gravitational field will have to expend energy to escape and, consequently, its spectral lines will be shifted toward the red. Since a red shift can also be produced by recession of the light source, this *gravitational red shift* can be isolated only when the recessional velocity can be determined independently. This can be done, for example, in the case of Sirius B, Procyon B, and 40 Eridani B, all of which are dwarf members of binary systems whose space motion can be evaluated from a study of the motions of the bright primary star. The observed gravitational red shift in the spectra of these stars (equivalent to 13 miles per second in excess of the known space velocity in the case of 40 Eridani B) confirms the prediction of the theory of relativity.

The nature of matter having a density a million times that of water, as well as the theory of the origin of white dwarfs, are our primary concern at this point. As we saw in Section 9.4, the conventional picture of the Bohr atom supposes a positively charged nucleus surrounded by negatively charged electrons. In this and all other models of the atom, most of the structure is empty space and only a tiny portion of the volume of the model is occupied by solid particles. In the gases that make up the interior of a white dwarf star, the atoms are almost completely ionized. When a nucleus has been stripped of its attendant electrons, the nucleus and the free electrons can be packed into a space that is almost infinitesimal compared with the volume occupied by a normal, un-ionized atom. There is still empty space in the atoms, but the density of the material will be incredible.

There is a minimum volume, depending on internal conditions in the star, to which a white dwarf can contract. When it reaches this point, it consists largely of a *degenerate gas* (a gas in which the electron shells around the nuclei have degenerated, or broken down). At this stage, the center of the star is relatively cool, but its surface temperature may be very high. The energy we receive from the star comes from the thin layer of hot gas at the surface.

The history of a star before it becomes a white dwarf was treated in the preceding section. Not all stars end up as white dwarfs, of course. Stars with a mass above a certain limit, called the *Chandrasekhar limit* (the upper limit on the mass of a stable white dwarf star, equal to approximately 1.2 times the mass of the sun) after the scientist who calculated it, take a rather different evolutionary turn.

## 11.8 PULSARS

If a degenerating star approaching the end of its evolutionary history exceeds the Chandrasekhar limit, densities greater than 100 million times that of water can be reached. As the density of stellar material increases, electrons are forced onto protons, and *neutrons* are formed. These particles have about the same mass as the protons, but are electrically neutral. The atomic particles in the star are now even more closely packed. A tablespoon of matter from such a star would weigh several billion tons on the earth. Internal pressure increases, and if the mass of the star does not exceed a certain limit, contraction stops. It is believed that the physical conditions necessary for reaching such densities are found in the centers of stars undergoing supernova explosions. The result is a *neutron star*, and in some cases, a particular type of neutron star, a *pulsar*. A neutron star has a diameter of 6 to 12 miles and a mass about 0.3 to 2 times the sun's mass.

The name *pulsar* for neutron stars whose electromagnetic output varies is probably unfortunate, since such stars do not actually pulsate. Their period of variation is incredibly short, ranging from 0.333 second (the period of the pulsar in the Crab nebula, shown in Figure 11.8) to almost 4.0 seconds, but this is due to their rapid rate of rotation.

All stars lose matter during their lifetimes, some quite violently as novas or supernovas. Many rotate and have magnetic fields. As evolving massive stars contract, the need to conserve angular momentum decreases their period of rotation and intensifies their magnetic field, which is "locked into" the stellar material and contracts with the star. The period of rotation can become very short if the star contracts sufficiently. The magnetic field of up to one trillion G rotates with the star, although its axis of symmetry may not coincide with the axis of rotation. This, in interaction with the surrounding medium, produces the observed variation in radio signals (electromagnetic radiation) from the star.

Of course, the incredible temperatures, densities, and magnetic fields of the neutron stars cannot be duplicated in earthly laboratories. The explanations and speculations about their apparent behavior and history are derived from mathematical calculations performed by high-speed computers and represent theoretical work of the most advanced nature. The discovery of the first pulsar was announced early in 1968, and about 60 are known today. With such

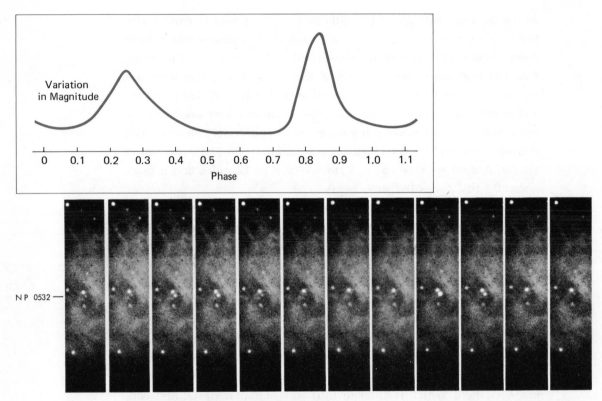

**Figure 11.8  The Pulsar in the Crab Nebula**
The period of this pulsar is 0.033 second. The curve (top) shows the variation in radiation with two maxima in each phase. The star (bottom) is indicated by the mark and by "NP 0532" at the left. The variation is evident. Kitt Peak National Observatory photograph by H. Y. Chiu, C. R. Lynds, and S. P. Maran.

a short time for research and such a paucity of examples, it is truly a remarkable commentary on the power of research in this area that so much has been accomplished.

## 11.9 COLLAPSARS

The point at which a neutron star stops collapsing is a critical point in its evolution. If its mass at this time exceeds 2 solar masses, the process of collapse will not stop, with astonishing results. The star will become so small (4 miles or less in diameter) and so dense, and its gravitational attraction so strong, that even light will be unable to escape. Such a star is called a *collapsar*, or *collapsed star*, and produces the phenomenon known as a *black hole*. A collapsar has practically zero volume and practically infinite density. It is black, since no light can escape from its surface, and any material that enters the region of space surrounding it will be swept into it and disappear. We can explain the behavior of ordinary stars, white dwarfs, and even pulsars in terms of Newtonian physics, but to understand the phenomena associated with a black hole we must turn to Einstein's general theory of relativity.

No collapsar has yet been detected. How, then, should we go about looking for evidence of their existence? Since no electromagnetic radiation can escape, direct observation of an isolated black hole is impossible. With luck, however, a black hole's effect on other bodies may be detected. Hence astronomers look for

1. a black hole that is part of a binary system and affects the motion of a companion (normal) star;
2. a black hole close enough to a normal star to draw in matter from it;
3. a black hole imbedded in a normal star; or
4. a black hole moving through a cloud of dispersed matter.

So new are the ideas and tentative conclusions in this area that future developments are impossible to predict.

## 11.10 THE EVOLUTION OF CLUSTERS

In H-R diagrams constructed for globular clusters, the main sequence extends upward to the left, as in Figure 11.9, to a point to the right of about +4 on the absolute magnitude scale. It terminates here, but branches upward to the right to the red-giant

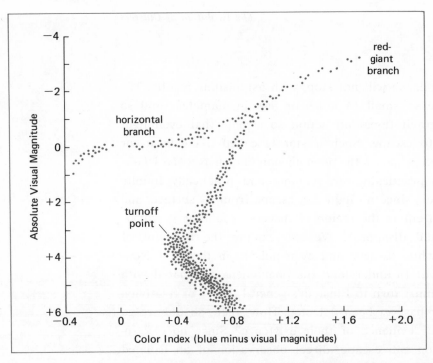

**Figure 11.9  An H-R Diagram for a Hypothetical Globular Cluster**
The main sequence is intact up to the turnoff point at a star of somewhat more than
the sun's mass. The turnoff point represents the star's position when it leaves the
main sequence and moves through the red-giant branch to the tip in the upper right,
at which time it may have expanded to 100 times the radius of the sun. It then
drops back, perhaps crossing the instability strip several times, and takes a position
along the horizontal branch. All stars in a globular cluster are believed to be
nearly the same age. They are distributed on the curve because some evolve faster
than others.

area of the diagram. The stars in the tip of the red-giant branch
have a temperature in the neighborhood of 4000°K and a magnitude
of −3 or −4. There is also a horizontal branch to the high-
temperature left edge of the diagram at an absolute magnitude of
about zero. The white dwarfs and the stars that would ordinarily
form the lower part of the main sequence in an H-R diagram do
not appear, since stars of this luminosity are not visible at the
distance of the clusters (although their existence may be inferred).

The point at which the main sequence of a cluster terminates
and branches to the right depends on the age of the cluster. In a
young cluster, the main sequence should be complete; but as the
cluster ages, the turnoff point moves down the main sequence

338

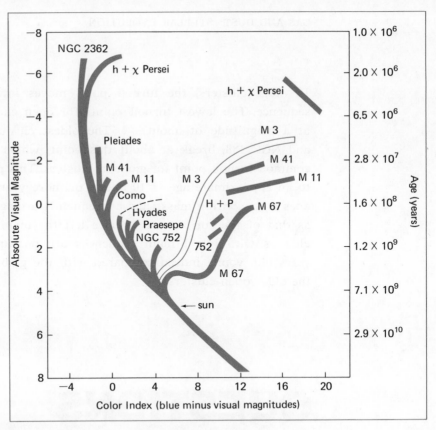

**Figure 11.10  A Color-Magnitude Diagram of Ten Galactic Clusters and One Globular Cluster (M3)**

The age of each cluster is obtained by moving from the turnoff point from the main sequence to the right side of the diagram. The magnitudes on the left edge are absolute visual magnitudes, and the B — V color index along the bottom is equivalent to temperatures of approximately 12,500° at the left edge and about 3000° at the right. (Adapted from a diagram by Allan R. Sandage).

until, theoretically, all the stars move to the right into the red-giant branch. Because the main sequence branches at about the same point in all the globular-cluster H-R diagrams, we assume that they are approximately the same age, perhaps 12 billion years or so. They are the oldest type of cluster, and their H-R diagrams are more or less similar to those of any system of extreme Population II stars.

H-R diagrams of open clusters are not so uniform, as Figure 11.10 shows. In some of the younger clusters, the main sequence is complete and very few stars have reached the red-giant branch.

In older clusters, the turnoff point moves lower on the main sequence. The lowest turnoff points for open clusters seem to be at a magnitude of about +4. The oldest clusters, such as M67 and NGC 188, break at about this point, which is also near the common turnoff point for globular clusters. This point corresponds to an approximate age of 10 billion or more years. In no instance does a star with a mass (and consequently a magnitude) the same as that of the sun appear to have left the main sequence. Those clusters with intact main sequences are perhaps a few million years old, young indeed compared with the globular clusters and the older open clusters.

# QUESTIONS

(1) What is the usual source of energy in an emission nebula? (Section 11.1)

(2) How is a reflection nebula different from an emission nebula? (Section 11.2)

(3) How do we know of the existence of dark nebulas? (Section 11.3)

(4) What is color excess? What causes it, and why is it important? (Section 11.4)

(5) What is an H II region? (Section 11.4)

(6) Why do we trace the evolution of a star by means of an H-R diagram? (Section 11.6)

(7) What is a Hayashi line? (Section 11.6)

(8) Briefly describe the evolution of a star similar to the sun. (Section 11.6)

(9) What is a white dwarf? (Section 11.7)

(10) What is the Chandrasekhar limit? Why is it important?
(Section 11.7)

(11) What is a neutron star? (Section 11.8)

(12) Describe three possible end-products of stellar evolution.
(Section 11.9)

(13) What happens to the turnoff point of a cluster as the cluster ages? (Section 11.10)

# CHAPTER 12
# THE MILKY WAY

Anyone who has gone out of doors on a clear, moonless night far from urban lights and smoke is at least casually familiar with the appearance of the pearly band of light across the sky called the Milky Way. The Milky Way is the visual evidence of the starry aggregation astronomers call the galaxy. The words *galaxy* and *galactic* have the same Greek root as *lactose,* meaning of or pertaining to milk.

Up to this point, every celestial body we have described, except some supernovas, has been a member of our galaxy. Galileo was the first to identify the starry composition of the Galaxy, but until the work of William Herschel in the second half of the eighteenth century no truly systematic investigation of it was carried out. The galaxy has a total mass about equivalent to that of 10 to 100 billion stars like our sun. Much of this mass is in the form of stars, but considerable quantities of gas and some dust are also present. The shape of the galaxy is that of a lenticular disk with spiral arms. The disk is about 30 kpc in diameter. Around it and concentric with it is a spheroidal region called the *halo,* or *corona.*

## 12.1 THE SIZE AND SHAPE OF THE GALAXY

Counting stars has, over the centuries, provided astronomers with various clues to the size and shape of the galaxy. It suggested, for one thing, that the galaxy was not infinite, that the number of stars thinned out as one moved further into space. The line of reasoning was as follows.

Since the apparent brightness of a star varies inversely with the square of its distance, there should be a statistically predictable increase in the average distance of stars of increasing magnitude (decreasing brightness). Suppose that, in order to obtain a systematic census of stars, we begin by counting all the stars with a magnitude of +1 or brighter. Discounting for the moment differences between absolute and apparent magnitudes, we can now assume that we have counted all the unobscured objects within a given radius of the earth. If we now extend our census to include stars with a magnitude of +2, we also, presumably, extend the radius of our search. A difference of 1 magnitude between two stars indicates a decrease in brightness by a factor of about 2.5. Hence by extending our census to include stars 1 magnitude fainter, we are extending the radius of our search by a factor of about 1.6 (the square root, approximately, of 2.5). Since we are counting stars in all directions and the volume of a sphere depends on the cube of the radius, the volume of space included in our census should increase by a factor of about 4 (the cube, approximately, of 1.6). If stars are uniformly distributed through space, the number counted when we extend our census to include those 1 magnitude fainter should also increase by a factor of about 4. Since this is not the case, we conclude that the stars thin out as one moves further into space and that there is a limit to the stellar system in which we live.

The line of reasoning followed in ascertaining the general shape of the galaxy was analogous to the deductions that might be reached by a man walking through a grove of trees. If he looks in a particular direction and sees many trees, apparently densely packed together, he is likely to conclude that the center of the

**Figure 12.1  The Milky Way**
These photographs form a mosaic of the Milky Way from Sagittarius to Cassiopeia.
Photographs from the Hale Observatories.

wood lies in that direction. Conversely, if trees appear sparse in another direction, he is likely to conclude that it is the shortest way to the edge. The same principle holds in studying the stars. When Herschel pointed his telescope in various directions in the sky and counted the number of stars in his field of vision, he found that they were most numerous near the Milky Way (Figure 12.1), whose central line we have come to call the *galactic plane.* Ninety degrees removed from the galactic plane, at the *galactic poles,* stars were few and far between. The concentration of stars, particularly faint ones, near the Milky Way indicated to him that the stellar system he was investigating extended farthest in the direction of the Milky Way and was least extensive at right angles to its plane.

Years of work counting stars in all directions led Herschel to the first scientific representation of our galactic system. He concluded that the galaxy occupied a disk-shaped space whose diameter was about five times its thickness and that the sun was at the center of the disk. Because of the dark rift visible in the Milky Way, he assumed that the disk-shaped region, or *grindstone,* as he called it, was split, sort of a cloven grindstone. He failed to realize that the dark rift was due not to an absence of stars, but to

344

clouds of interstellar material. Stars are actually numerous in the direction of the rift, but they are obscured by absorbing material. Herschel suggested no other objects outside our galaxy, since he had no reason to suppose there were any.

Following the era of Herschel and his immediate successors, there was a period in which astronomers were more interested in other features of the universe than in the structure of the galaxy. About the beginning of the twentieth century, interest in galactic astronomy revived, and the Dutch astronomer Jacobus Kapteyn began a model of the galaxy that showed it to be about 8 kpc in diameter and about 2 kpc thick.

In 1917, Harlow Shapley, who was investigating the distances and directions of the globular clusters, monumentally increased the suggested size of the galaxy to a diameter of about 100 kpc and a thickness of approximately 10 kpc. Shapley based his estimate on an examination of RR Lyrae stars, which are quite numerous in clusters and which he believed to have a mean absolute magnitude of about zero. Actually, they are somewhat fainter. We now estimate their mean absolute magnitude to be about +0.6. This makes the clusters somewhat closer than Shapley thought them to be. For example, the globular cluster in Hercules, M13, is now believed to be approximately 8 kpc away, not 11 or 12. Nevertheless, Shapley's work represented a pioneering step in our understanding of the galactic system.

## 12.2 DETAILS OF GALACTIC STRUCTURE

In 1924, Edwin Hubble established that the spiral nebulas, as they were then called, were outside the Milky Way. Larger telescopes permitted the resolution of these extragalactic systems into stars, and most astronomers concluded by analogy that our galaxy too was a spiral. Thus a provisional understanding of the structure of the galaxy began to emerge. Shapley had postulated a spherical system of

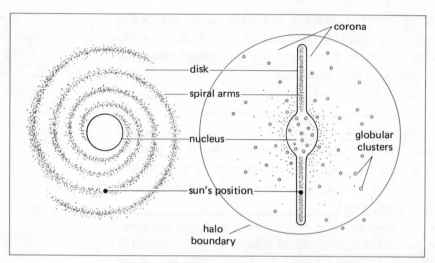

**Figure 12.2  A Schematic Representation of the Milky Way**
The exterior diameter of the spiral is about 30 kpc. The globular clusters
and some stars are found in the halo outside the disk,
but the entire structure is concentric to the nucleus of the galaxy.
Compare with Figure 12.8.

globular clusters. Astronomers now assumed that the galaxy was
a large flattened disk concentric with Shapley's cluster system.
Early estimates of its diameter ranged as high as 100 kpc. Today, a
provisional figure of approximately 30 kpc is considered acceptable.

As Figure 12.2 shows, the Milky Way consists, first, of a nucleus
similar to that observed in the Andromeda spiral M31 and other
galaxies of the same type. Stars in the nucleus are closer together
than stars in the sun's neighborhood. Extending outward from the
nucleus are spiral arms whose appearance presumably resembles
closely the arms of similar spirals observed from the earth. The
rotation of the galaxy causes the spiral arms to wrap about the nucleus. Above and below the disk (which contains the nucleus, the
arms, and other objects in the same general region), is the corona,
or halo. This is a spheroidal volume much less densely populated
than the disk and mapped by Shapley in his study of globular clusters. Objects in the halo do not necessarily rotate with the rest of

the galaxy, but move individually in highly eccentric and inclined orbits about the center of mass represented by the center of the nucleus. The halo, a very old region, can be said to be associated with the disk, but it is not an integral part of it. Clusters and RR Lyrae stars have been detected from 30 to 50 kpc above and below the galactic plane, but no H II regions exist in the halo.

In the 1940s, Walter Baade discovered luminous stars in the arms of the Andromeda spiral M31. In 1951, astronomers at the Yerkes Observatory discovered Class O and B stars and emission nebulas of a similar nature in the Milky Way. Observations of any objects near the galactic plane must be regarded with extreme caution, since the high absorption of light by interstellar material makes them appear fainter than they are. However, when an appropriate allowance for absorption was made, it became evident that, just as in the exterior spirals, O and B stars were associated with emission nebulas. The conclusion that these features outlined the spiral arms of our galaxy led to the first optical tracing of these arms. The sun is in what is now called the *Cygnus arm*. This arm can be traced, by means of the O and B stars and their associated emission nebulas, through about 200° of galactic longtitude—that is, slightly more than halfway around the sky. The arm is roughly circular, with its presumed center in the galactic nucleus; its radius decreases by about 1.5 kpc as it is traced clockwise around the galaxy.

Outside the Cygnus arm and concentric with it is another arm called the *Perseus arm* (because it includes the double cluster in the constellation of Perseus). It has been traced through approximately 100° of galactic longitude and at a radius of about 10 kpc from the provisional location of the galactic center. Inside the Cygnus arm is the Sagittarius arm. Its radius is 1 or 2 kpc less than that of the Cygnus arm, but its center is evidently in the galactic nucleus. About 3 kpc from the center is an expanding arm. All these arms appear to move clockwise as seen from the north. There is some disagreement among astronomers about details, but the basic conclusion seems to be that our galaxy resembles M31 in Andromeda. It is more compact than a typical loose spiral.

Although the mass of interstellar material in the sun's vicinity approximates the mass of the stars, only 2 percent of the total mass of the galaxy is estimated to consist of gas and dust, while the rest is in the form of stars. The stars in the arms of the spiral are a younger population, and interstellar gas and dust represents the birthplace of the stars. This is the current picture of the galactic spiral as determined from an investigation of the bright Class O and B stars and their associated emission nebulae. It is in good agreement with the general picture that we get from observations of extragalactic spirals and in reasonable agreement with the findings of radio astronomers.

Investigation indicates that the physical center of the galaxy is in the general direction of the constellation of Sagittarius, which has been assigned a galactic longtitude of 0°, and is approximately 8 or 9 kpc from the sun. Figures 12.3 and 12.4 show parts of this region of the galaxy.

## 12.3 THE MOTION OF THE MILKY WAY

The galaxy is able to maintain its form and avoid collapsing into the central nucleus because it rotates about its center. Kapteyn, while investigating selected areas of the sky, discovered what he called *star streaming*, an apparent preference on the part of stellar objects for movement in two directions. This preference could not be accounted for by the orbital motion of the earth about the sun and was, in fact, early evidence of galactic rotation.

Celestial objects can be classified into two general categories according to their motion: *low-velocity objects*, which have velocities with respect to the sun of 5 to 30 miles per second, and *high-velocity objects*, with velocities with respect to the sun in excess of 50 miles per second. However, if the motion of these two classes of

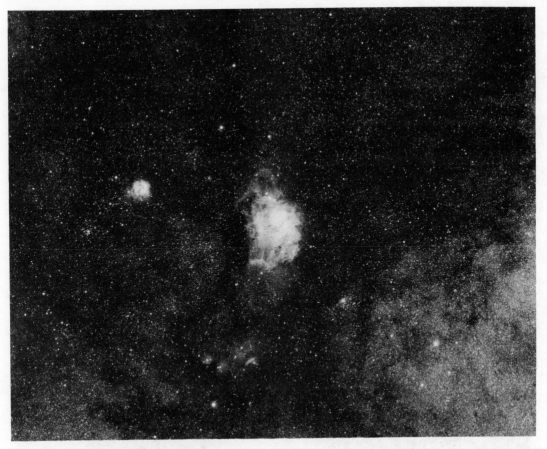

**Figure 12.3  A Portion of the Milky Way**
This view of the Milky Way in Sagittarius shows two large nebulas, NGC 6514 and NGC 6523, and several galactic clusters. Photograph from the Hale Observatories.

objects is considered not with respect to the sun, but with respect to the general structure of the galaxy, it is the "high-velocity" objects that have a low velocity. The motion of the sun due to the rotation of the galaxy causes objects that are not revolving as rapidly as the sun to appear to have a high velocity with respect to it.

The sun is provisionally believed to be about 8 or 9 kpc from the galactic center and to have a rotational velocity about the center of 150 miles per second. Evidence of the rotation of the galaxy is provided by differential effects measurable in terms of the radial velocities of objects with reference to the sun. Figure 12.5 shows how the radial velocities obtained from measuring Doppler shifts

349

Figure 12.4 **Star Clouds**
This photograph shows an area of the Milky Way near that in Figure 12.3. This is the
brightest part of the Milky Way, not far from the assumed direction of the center
of our galaxy. Photograph from the Hale Observatories.

give us information about the rotation of the galaxy. Stars with a
galactic longitude of 45° or 225° have a recessional average radial
velocity. Those with longitudes of 135° and 315° have approach velocities. Those with longitudes of 0°, 90°, 180°, and 270° have an average velocity of approach or recession of zero.

350

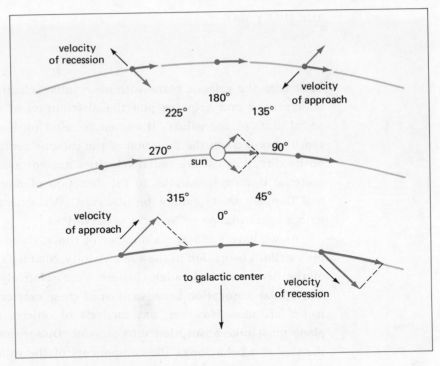

**Figure 12.5 The Effect of the Rotation of the Galaxy on Radial Velocities**
The arrows parallel to the direction of rotation of the galaxy indicate approximately,
by their direction, the direction of the stellar motions and, by their length, the
relative speeds of the stars. Stars in longitudes of 0°, 90°, 180°, and 270° moving
parallel to the sun's motion have no preferential radial velocities. Stars with
longitudes of 45° are leaving the sun behind and those in longitude 225° are being
left behind by the sun, since recessional velocities are dominant in these two
directions. On the other hand, the stars in longitude 135° are being overtaken by the
sun and those in longitude 315° are overtaking the sun. These stars have velocities
of approch.

The Dutch astronomer J. Oort has related the observed velocities of galactic objects outside the nucleus to their distance by the
following formula:

$$V = rA \sin 2\,l$$

where $V$ is the velocity of the object, $r$ is its distance from the sun,
$A$ is a constant, and $l$ is the object's galactic longitude. An analysis
of stellar radial velocities shows $A$ to be 11.6 miles per second per
kpc. Thus one need only observe the longitude and velocity of approach or recession of an object to find $r$, its distance from the sun.

This formula is effective in determining the distance of objects

351

in or near the galactic plane with observable velocities of recession or approach and helps us plot the distribution of material in the spiral arms of the galaxy. It cannot be used to determine the distance of objects in the direction of the galactic center, or in the opposite direction (where radial velocities are not available since the material is moving parallel to the direction of motion of the sun and Doppler shifts cannot be observed). Velocities are measured with a spectroscope or by radio observations.

As we have remarked a number of times, there is considerable interstellar absorption in the sun's vicinity. Shapley's measurements of the distance of globular clusters were relatively unaffected by interstellar absorption because most of these clusters have high galactic latitudes. However, any analysis of objects in the galactic plane must take absorption into account. Disagreements regarding the size and behavior of the components of the galaxy can usually be traced to different interpretations and assumptions regarding absorption by interstellar material and consequent differences in interpretations of luminosity parallaxes.

To summarize, our present understanding of the rotation of the galaxy is as follows: The central region, which is more compact than the rest, rotates clockwise as seen from the north galactic pole, essentially as a solid body. Outside the central nucleus, which has a radius of some 3 kpc, the rotation approximates what is known as a Keplerian motion, similar to that of the planets of the solar system. The more distant stars revolve about the center more slowly and have longer periods than the closer ones. Figure 12.6 shows the change in the galaxy's orbital velocity as the distance from the center increases. The sun has a velocity of revolution in the neighborhood of 150 miles per second and takes approximately 220 million years to make one revolution. This period of time has been referred to as a *cosmic*, or *galactic*, *year*. Since its birth (about 5 billion years ago), the sun has completed 20 to 25 revolutions. Considerable mystery is attached to the permanence of the galaxy's spiral structure over such a long period, and some unknown agency

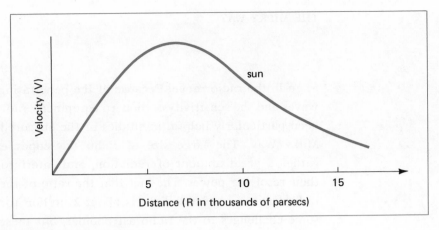

Figure 12.6 **The Orbital Velocity of the Galaxy**
**as a Function of the Distance from the Center**
From the nucleus to about 5 kpc from the center, the galaxy rotates essentially
as a solid. Farther out, a motion resembling the Keplerian velocities of the
planets predominates and the outer edges rotate slowly. If the galaxy were a solid
disk, the velocity of rotation would increase steadily with the distance from the
center and the graph would be a straight line.

may be operating to preserve it. Because spiral structure is most
evident when one looks at very young stars (and associated emission regions), it may be that spiral structure is not permanent but
is continually refreshed and recreated by the forces responsible for
star formation.

Objects in the halo, both globular clusters and individual stars,
do not share the general rotation of the disk population, but move
in orbits about the center that resemble the orbits of comets about
the center of the solar system. Sharing the rotation of the galactic
nucleus is a gaseous ring, expanding outward in all directions from
the nucleus at about 30 miles per second. What impels this cloud of
gas outward from the center and how it has been replenished (if,
indeed, it has been replenished) over the billions of years that the
galaxy must have existed is not known.

## 12.4 THE RADIO PICTURE OF THE MILKY WAY

In Chapter 3, we discussed the electromagnetic spectrum and the
development of instruments used to measure extraterrestrial radi-

ation in the radio range. Because of the penetrating power of radio waves and the sensitivity of modern equipment, radio astronomy has been particularly helpful in studies of the obscured portions of the Milky Way. The large size of radio telescopes enables them to gather a great amount of radiation, and interferometry enhances their resolving power. The fact that the ratio of the wavelengths of radio waves and visible light is about 2 million to 1 has presented some challenges to the radio astronomer, but these have been successfully met. Since both radio waves and light are forms of electromagnetic radiation, they are subject to the same basic laws, and similar phenomena are associated with them. There is, for example, a continuum in the radio spectrum. There are absorption and emission "lines," and there are Doppler shifts similar in principal effects to those of visible light.

Two basic types of emanations are received by radio astronomers. The first is *continuum radiation,* analogous to the continuous spectrum the optical astronomer finds in the background of solar and stellar spectra. However, it is not definitive in the sense that particular features can be used to measure such things as Doppler effects. Early in the history of radio astronomy, a *contour map* was developed of the general region of the Milky Way (Figure 12.7). Telescopes, many of them movable only north and south along the meridian, were used to scan the Milky Way. The radiation received from the antennas was fed into an amplifier and then into a recorder. As the earth turned from west to east through 24 hours, a band completely encircling the sky at a particular north or south setting was scanned. At the end of that time, the antennas were moved slightly and another belt was scanned. Coordinating these observations gave astronomers a measure of the intensity of the continuum of radio noise in the area covered by the antennas. They soon discovered that the flux was much higher in the Milky Way than in other regions of the sky, although isolated radio sources outside the Milky Way were found as well.

The intensity of radiation from the Milky Way itself is not uni-

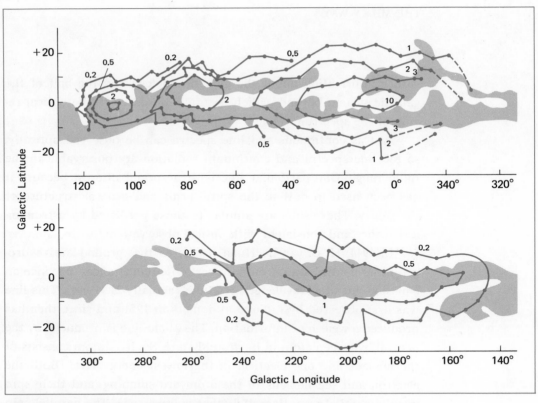

Figure 12.7 **An Early Contour Map of the Milky Way**
The shaded areas indicate the optical picture of the Milky Way, produced by
radiation in the visual range. The contour lines indicate equally intense radiation
from the Milky Way in the radio range. (Diagram from Reber's *Radio Data*.)

form. There is a peak in radio reception in the general direction of
what has been determined optically to be the galactic nucleus. Due
to the higher penetrating power of radio waves, this peak is relatively more prominent than the corresponding peak in visible radiation. Just as radio waves penetrate the walls of our dwellings,
so too they penetrate the obscuring dust and gas in and near the galactic plane. Hence, although the radio astronomer cannot "see"
with his instrument the way the optical astronomer does, he can
delineate, with precision, the concentrations of radio emitting regions at and near the galactic center. He thus supplements and extends the work of the optical astronomer, whose view is obstructed
by the obscuring medium. In fact, the radio astronomer is able to
penetrate regions that no optical astronomer has ever observed and

355

gather information about the galactic nucleus and the half of the galactic system beyond, which is totally obscured in the optical region of the spectrum.

Just as continuous and line spectra can be observed optically, so too line spectra and continuum radiation are observable in the radio range. One particular line is of a special interest because it has been used to outline the spiral arms and general structure of the galaxy. The results are similar to those predicted by astronomical theory and consistent with optical observations.

In 1944, H. C. van de Hulst predicted that ground-level hydrogen in its neutral state would, in certain circumstances, produce an emission line at a wavelength of 21 cm (actually 21.11 cm). This line was first observed by radio astronomers in 1951 and since then has produced a wealth of information. The 21-cm line is produced by the neutral hydrogen atom in its ground state. Such an atom consists of the nucleus and one electron in the lowest energy level. Both the electron and the nucleus in the atom are spinning, and their spin may be parallel or antiparallel (that is, reversed). The parallel spin represents a slightly higher energy level than the antiparallel spin. If the axis of spin of the electron is reversed by absorption or emission of the requisite amount of energy, an absorption or emission line will be produced. This results in the appearance of the 21-cm line against the continuous background of the radio noise in the galactic plane. The Doppler effect makes the velocity of the hydrogen clouds producing the line measurable.

Using the 21-cm line to trace the structure of the galaxy has a number of advantages. First, H I regions (regions of neutral hydrogen), unlike H II regions (regions of ionized hydrogen), are not visible in a simple optical telescope. Second, the radio waves at 21 cm are highly penetrating, as we noted above. Consequently, the shape of the galaxy can be plotted despite the clouds of gas and dust that are opaque to other types of radiation, including visible light.

Figure 12.8 shows the results of a radio trace of the Milky Way. Radio astronomers, notably those in the Netherlands plotting the

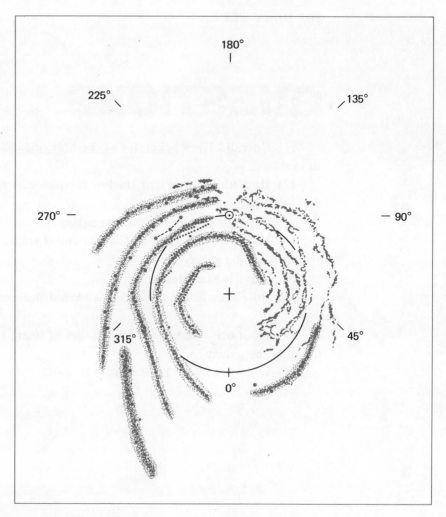

**Figure 12.8 The Spiral Structure of the Galaxy**
This diagram is a tracing of the radio pattern from Leiden and Sydney. The numbers indicate galactic longitude. The sun is indicated by a circle and the center of the galaxy by a cross.

northern part of the sky and those in Australia plotting the southern portion, have provided remarkable maps of the distribution of H I regions in our galaxy. Such maps, along with the earlier optical observations, have been extremely valuable in developing a clear picture of the galaxy in which we live.

# QUESTIONS

(1) How did Herschel arrive at the first scientific description of the galaxy? (Section 12.1)

(2) How did the work of Harlow Shapley contribute to our understanding of the galaxy? (Section 12.1)

(3) Describe the rotation of the galaxy. (Section 12.2)

(4) How does the study of Class O and B stars tell us about the structure of the galaxy? (Section 12.2)

(5) What is star streaming? (Section 12.3)

(6) How long is the cosmic year? What determines its length? (Section 12.3)

(7) What are some of the advantages of using radio astronomy to study the galaxy? (Section 12.4)

(8) What is a 21-cm line? How is it used? (Section 12.4)

# CHAPTER 13

# GALAXIES AND QUASARS

We have discussed the fact that the early catalogs of nebulas included a variety of objects. The term *nebula* was used to describe almost any fuzzy patch of light in the sky. True nebulas (clouds of gas) illuminated by enmeshed stars were included in the catalogs, but so were other objects such as star clusters and what later proved to be exterior galactic systems. Even before 1920, it had been recognized that many of the so-called nebulas were spirals, and there was considerable discussion about their true nature and distance. We shall discuss these extragalactic systems, and the puzzling objects called quasars that look like stars but behave like galaxies, in this chapter.

## 13.1 THE RELATIONSHIP OF THE SPIRALS TO OUR GALAXY

In 1920, a debate was held between Harlow Shapley and H. D. Curtis regarding the nature and distance of the spirals. Shapley maintained that they were within or near the Milky Way; Curtis took the view that they were extragalactic. The idea of other galaxies was not new. The philosopher Immanuel Kant proposed the concept of "island universes" in the middle of the eighteenth century. Strictly speaking, the word universe has no plural, but this phrase became a catchword. It meant, however, in this instance, other galaxies outside our own, and this was the usage accepted.

Curtis argued that some of the galaxies, such as M31 in the constellation of Andromeda (Figure 13.1), were as much as 150 kpc distant. He based his argument on the comparison of novas observed in M31 and novas a known distance away in our own galaxy. He also argued that since the spirals varied greatly in size and apparent brightness, they would have to be at very unequal distances for their intrinsic brightness to be at all similar. The superficial distribution of the galaxies did lend some credence to the idea that they might be associated with the Milky Way, for they seemed to avoid the central plane of the Milky Way and to be arranged symmetrically on either side of it. They do not, however, form a system concentric with the Milky Way. They appear to be more numerous in high galactic latitudes than in the direction of the galactic plane because the distribution of absorbing material produces a *zone of avoidance* (Figure 13.2), not because of any real relationship to our galaxy.

One mysterious property a large number of spirals seem to possess is sizable radial velocities, almost all recessional and greater than the radial velocities of any other class of objects. At one time, it was suggested that a repulsive force unknown locally but emanating from our galaxy operated on a cosmic scale, driving the spirals further away.

The arguments over the nature of the spirals are strangely reminiscent of the conflicting theories held in the early days of the study of Cepheid variables. In this case, however, spurious proper

Figure 13.1 **The Galaxy M31 in Andromeda**
This photograph shows NGC 224 (M31) in Andromeda and two satellite galaxies,
NGC 205 and NGC 221. Photograph from the Hale Observatories.

motions also confused the issue. Proper motions are normally large
for nearby objects and small for distant ones. In the 1920s and before, the figures for the proper motions of what later proved to be
extragalactic objects were considerably overestimated. The mistakes
resulted from observational errors, but since there was no way astronomers could know this at the time, they became an integral part
of the controversy.

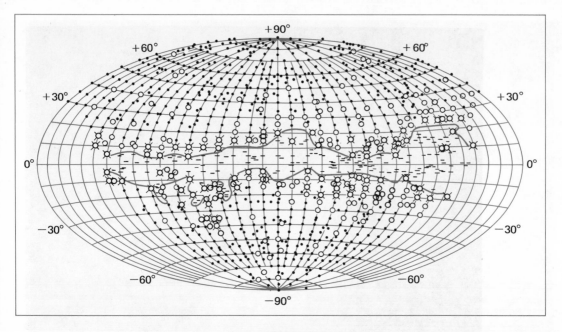

Figure 13.2 **The Zone of Avoidance**
Obscuring material in the plane of our galaxy prevents an unobstructed view of the galaxies outside our own. The areas in which the interstellar medium is so dense that no exterior galaxies can be seen is called the zone of avoidance. The size of each dot indicates the number of galaxies counted with the 100-inch telescope. Dashes and open circles indicate few or no galaxies. The zone of avoidance is irregular but coincides roughly with the visible Milky Way.

In 1924, Edwin Hubble settled the argument once and for all. Using the 100-inch telescope on Mt. Wilson in California, he discovered (and eventually produced light curves for) a number of Cepheid variables in M31. Centering the field of view of the huge telescope on the outer regions of M31, he was able to resolve them into individual stars (Figure 13.3). Hubble's results depended on the following hypotheses: First, he assumed that the variables were in the nebula, or spiral, and not in the foreground. As more variables were discovered in similar positions, the probability of there being stars in the foreground became negligible. Second, because of a lack of contrary evidence, he assumed that there was no absorption between the earth and M31. His work preceded by some six years that of Robert Trumpler on the absorbing material in the Milky Way. The Andromeda spiral is close to the galactic plane but has a high enough galactic latitude so that light reaching the earth encounters

362

Figure 13.3  **An Enlarged Section of M31**
This photograph shows the resolution of the south preceding region into stars by a 100-inch telescope. Photograph from the Hale Observatories.

only a moderate amount of gas and dust. Hubble's third assumption was that the Cepheid variable stars in the Andromeda spiral varied in the same way as those in our own galactic system.

The distance of the Andromeda spiral, according to Hubble's measurements, was about 450 kpc. Since the maximum radius of Shapley's sphere of globular clusters was 150 kpc, or about one-third the distance measured by Hubble, it was definitely settled that M31 was not within the Milky Way. In the years following Hubble's investigations, the distance to the Andromeda spiral was found to be four or five times as great. Estimates of the diameter of the Milky Way, meanwhile, were reduced, placing the Andromeda spiral well outside our galaxy. From a cosmological standpoint, however, M31 is very close and is a member of what we call the Local Group of galactic systems.

363

## 13.2 A DISTANCE SCALE FOR GALAXIES

Of the billions of galaxies in the universe, only about 30 are close enough for the behavior of Cepheid variables to be observed, and Cepheids have not been detected in all 30. However, these 30 provide the foundation for constructing a distance scale based on similarities between stars and galaxies of certain types. We can find the mean absolute magnitude, and hence the distance, of Cepheid variables by observing their periods. This method is effective to a distance of about 6 mpc. Beyond this point, estimates of the distances of galaxies are based on observations of other, brighter stars. In each galaxy, there are stars more brilliant than the brightest Cepheids. An empirical upper limit is established for their absolute magnitude, based on observations of the brightest stars in galaxies whose distances are known from observations of Cepheid variables. Once the stars' approximate true magnitude is known, of course, their distance can be calculated. The distances of galaxies too far away to permit the resolution of even the brightest individual stars are calculated from estimates of their total luminosity, based on observations of similar nearby galaxies. Distances can also be estimated, in some cases, from supernovas (Figure 13.4), which have a maximum absolute magnitude in excess of $-16$. A star this bright can be observed from very far away and further extends the distance scale.

Extragalactic objects have red Doppler shifts in their spectral lines greater than those of any other class of objects. This predominance of red shifts indicates that they are receding from, rather than approaching, the Milky Way. When Hubble had determined the location of enough galaxies to support a generalization, he announced that the Doppler red shift also provided a way of measuring their distances. He concluded that all the galaxies, except a very few nearby ones, were receding from the earth and our own galaxy and that the shift of their spectral lines was proportional to their distance from us.

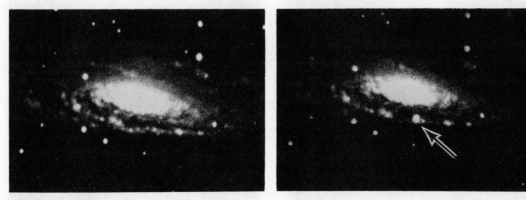

**Figure 13.4  A Supernova in a Galaxy**
These photos show two views of NGC 7331, before and during the maximum of a 1959 supernova. Lick Observatory photographs.

This last discovery gave rise to the theory of the expanding universe, which we shall discuss later in the chapter. As telescopes became more powerful and techniques more refined, greater red shifts were observed. Many astronomers found considerable psychological difficulty in accepting these findings and the notion of an expanding universe. No other satisfactory explanation, however, has yet been offered, and at present the idea of an expanding universe seems to be well established.

## 13.3 CLASSIFYING EXTRAGALACTIC SYSTEMS

As the number of galaxies cataloged proliferated, some sort of classification system became necessary. Hubble was the first to devise such a system. He divided the galaxies into four basic types: *spiral*, *barred spiral*, *elliptical*, and *irregular*.

The spirals are the most noticeable, but probably not the most numerous, of the galaxies. Their average luminosity is high, and their form sufficiently distinctive, to make them easily recognizable.

Hubble classified the ordinary spirals (Figure 13.5) on the basis

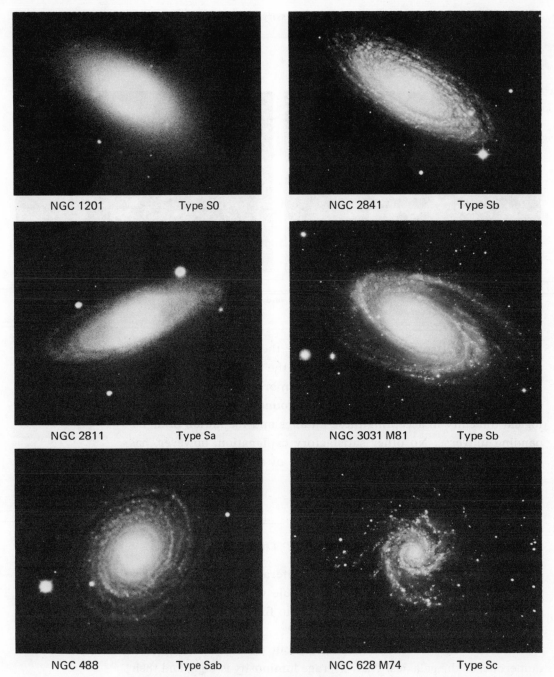

NGC 1201          Type S0

NGC 2841          Type Sb

NGC 2811          Type Sa

NGC 3031 M81      Type Sb

NGC 488           Type Sab

NGC 628 M74       Type Sc

Figure 13.5 **Some Examples of Normal Spirals**
These are normal spirals, S0 to Sc. Compare with Figure 13.8. Photograph from the Hale Observatories.

366

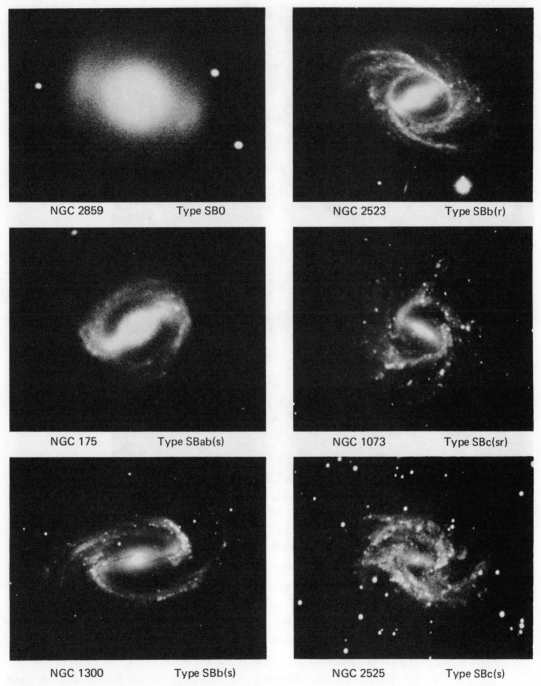

| NGC 2859 | Type SB0 | NGC 2523 | Type SBb(r) |
| NGC 175 | Type SBab(s) | NGC 1073 | Type SBc(sr) |
| NGC 1300 | Type SBb(s) | NGC 2525 | Type SBc(s) |

Figure 13.6 **Some Examples of Barred Spirals**
Compare with Figures 13.5 and 13.8. Photograph from the Hale Observatories.

Figure 13.7 **The Barred Spiral NGC 7479**
Note the transverse bar through the nucleus. Hale Observatory photograph.

of their appearance, as Sa, Sb, or Sc galaxies. An *Sc galaxy* does not have a prominent nucleus. The arms are rather loosely coiled and seem to contain a large proportion of the total material of the system. An *Sb spiral*, of which M31 (Figure 13.1) is a good example, has a more prominent nucleus and the arms are more closely wrapped about the center of the system. An *Sa spiral* has a very prominent nucleus and tightly coiled arms.

Barred spirals (Figures 13.6 and 13.7) are far less numerous than ordinary spiral galaxies. In an ordinary spiral, the arms emerge

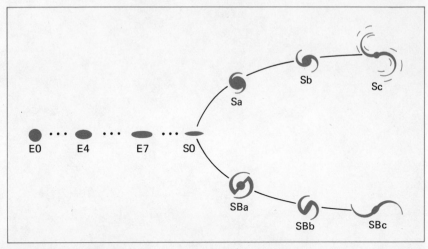

Figure 13.8 **Hubble's Classification of Galaxies**
The S0 class was a later addition by Hubble to the classification system. Although it has been suggested that this is an evolutionary sequence, Hubble himself did not say so.

from the nucleus and the entire galaxy resembles a pinwheel. A barred spiral, however, has a bar, or luminous arm, lying across the nucleus. At the end of the bar, which may be several times the diameter of the nucleus in length, the spiral arms break off at right angles. Hubble also divided the barred spirals into three classes according to their appearance: SBa, SBb, and SBc. The SBa galaxies, like the Sa spirals, have the most tightly wound arms while in the SBb and SBc galaxies the arms are more loosely coiled.

Figure 13.8 is an adaptation of Hubble's classification system. He did not suggest that this arrangement was an evolutionary sequence, but some astronomers have interpreted it in that way. The S0 galaxies did not appear in the original diagram, but were added at a later date.

For some time, the spiral galaxies were thought to be the most typical and the most numerous of the exterior systems. It now appears, however, that over half the exterior galaxies may be elliptical. There are eight classes of elliptical galaxies, E0 through E7. The

369

Figure 13.9 **The Large Magellanic Cloud**
This is an irregular system near our galaxy. Lick Observatory photograph.

E0's are circular; the digits 1 through 7 indicate the ellipticity, or departure from circularity, of the remaining classes. Obviously, an E6 or E7 galaxy must be lenticular, but an E0 may be a flattened galaxy whose principal plane is parallel to the plane of the sky. The question thus arises, Does the scale E0 to E7 reflect the true shape of the galaxies, or is their apparent ellipticity an accident of orientation? It is easy to compute the proportions of each class that would result if all elliptical spirals were lenticular and their apparent shape a result of their orientation. When this is done, it is evident that some of the elliptical systems must be spherical, or very nearly so.

The two extragalactic systems with the greatest apparent brightness are irregular galaxies called the large Magellanic cloud (Figure 13.9) and the small Magellanic cloud. Visible in the Southern Hemisphere, they were named for the great Portuguese navigator, Magellan, although their existence had been recorded by travelers gen-

Figure 13.10 **An S0 galaxy with Peculiar Characteristics**
This is NGC 2685 in Ursa Major. Photograph from the Hale Observatories.

erations before Magellan's time. They are approximately 50 kpc from us and appear, as their name indicates, as two irregular cloudy or hazy patches in the southern sky. The large Magellanic cloud is populated by blue stars and luminous nebulas. There are many clusters, including a large number of globular ones. There is also an abundance of Class O and B stars and associated nebulas and Population I and II objects, although there is little or no dust in the small cloud.

The S0 galaxy (Figure 13.10), which did not appear in Hubble's original diagram, has been placed at the juncture of the two spiral branches and the beginning of the elliptical sequence. In the S0 galaxies, the arms have apparently disappeared completely and evidentally no stars are being formed. We shall see later in the chapter that S0 galaxies are very numerous in the rich regular clusters of galaxies.

A later and more complete system of classifying galaxies, developed by Morgan and Mayall, is based on the condensation of their centers and their spectral class. Since a galaxy consists of stars and, to a minor extent, emission nebulas, its spectrum depends on the type of star or stars that provide most of the light. Galactic spectra resemble stellar spectra and are classified in much the same way, using the same letters as the stellar spectral sequence. Morgan and Mayall divided the galaxies into seven groups. Group one contains the irregular galaxies. Groups two through six are spirals of increasing condensation and tighter arm structure. Group seven includes the ellipticals. We shall now proceed to an examination of the characteristics of each of the general types of extragalactic systems, beginning with the spirals.

## 13.4 CHARACTERISTICS OF SPIRAL GALAXIES

The Andromeda spiral M31, also designated NGC 224, is a Class Sb spiral in the constellation of Andromeda. It can be seen with the naked eye and appears as a faint hazy patch of light in a pair of binoculars. It is almost on edge as seen from the earth, being tilted 15° from the line of sight, or 75° from the plane of the sky. Like other spirals including the Milky Way, it rotates in such a way that the arms trail from the nucleus. Photographs of M31 clearly show lanes of obscuring gas and dust, particularly in the regions further away from the nucleus. Enmeshed in this material are bright Class O and B stars. Surrounding the system is a halo containing many globular clusters. M31 is probably larger than the average spiral, and it is certainly larger than the Milky Way. As far as astronomers can tell, the spirals seem to range in diameter from 7 to approximately 40 kpc.

From 1924 to 1930, estimates of the diameter of the Milky Way were reduced from approximately 100 kpc to about a third of that figure; later, they were reduced still more. In the 1930s, estimates of the distance to M31 were increased to about 300 kpc, thus doubling the corresponding estimates of its linear diameter. Originally, the diameter of M31 was thought to be approximately a tenth of the diameter of the Milky Way. Other galaxies observed were obviously smaller than M31, and this placed us in the cosmologically embarrassing position of living in what was by far the largest known galaxy. Reducing the diameter of the Milky Way and increasing the distance and consequent linear size of the Andromeda spiral did much to eliminate this embarrassment. In the 1950s, the scale of extragalactic distances was further increased, by a factor of two or more, thus placing M31 perhaps about 1 mpc from us and correspondingly increasing its size. Astronomers now believe that M31 is 20- or 25-percent larger in diameter than the Milky Way.

Barred spirals constitute a minority of the spiral population. Perhaps about one-third of all spirals are barred. As we have indicated, their chief distinguishing characteristic is a transverse bar through the nucleus, from whose extremities the arms extend at right angles. When they are seen exactly or nearly on edge, the barred spirals cannot be distinguished from normal spirals. This tends to diminish the number of barred spirals recorded.

The angle of tilt to the line of sight differs for various spirals. In all those seen edgewise (Figure 13.11), a dark rift of obscuring material is clearly evident. The spiral character is clearest in those parallel to the plane of the sky (Figure 13.12).

## 13.5 CHARACTERISTICS OF ELLIPTICAL GALAXIES

A good example of a large elliptical system is the enormous M87 (Figure 13.13), which has been assigned a provisional mass in the neighborhood of 10 trillion times the mass of the sun. It is rich in

Figure 13.11 **A Spiral on Edge**
This is NGC 4565 in Coma Berenices seen edge on. Photograph from the Hale Observatories.

Population II objects and stars, as are most elliptical galaxies. This is an old population.

Not all elliptical galaxies are as large as M87; indeed, it is certainly one of the largest. Some, like the dwarf NGC 147 in Casseopeia (Figure 13.14), are so small that they are little more than oversized globular clusters. Because many of the elliptical galaxies are inconspicuous and have a low intrinsic luminosity, their number tends to be underestimated. Probably, they are the more common type of galaxy.

The elliptical systems have no spiral character and give no evidence of rotational symmetry. However, they are resolvable into stars, although their distances make it impossible to evaluate the

374

**Figure 13.12  A Spiral in Ursa Major, M81**
This galaxy is also known as NGC 3031. Note the dark nebulosity in the arms.
Photograph from the Hale Observatories.

main sequences and find any turnoff points that would indicate their age. A lack of dust and gas and a predominance of Population II stars makes them, apparently, the oldest of the galaxies. It has been suggested that if the classical Hubble diagram is to be taken as an evolutionary track (which, as we have noted, Hubble himself never suggested), the sequence should be from Sc and SBc through the two spiral branches to S0 and then from E7 to E0. Of course, the elliptical galaxies would have to be classified objectively rather than according to their shapes as viewed from the earth. However, so

Figure 13.13 **The Elliptical Galaxy M87**
This is a long-exposure shot of M87 (NGC 4486). The fuzzy patches nearby are globular clusters. Lick Observatory photograph.

far as we know, there are no inherent differences among the elliptical systems other than the apparent variation in elongation.

## 13.6 CHARACTERISTICS OF IRREGULAR GALAXIES

Irregular systems have little in common except a lack of symmetry. They appear to be a small minority among extragalactic objects, but

**Figure 13.14  A Dwarf Elliptical Galaxy**
This dwarf elliptical system, NGC 147, is a member of the Local Group in the constellation of Cassiopeia. Photograph from the Hale Observatories.

if many are small and intrinsically faint, they may be more numerous than is generally believed. The irregular galaxies have been divided into two classes. The first, Irr I, abounds in Class O and B stars and nebulous material. Irr II, the second type, has not been

377

resolved into stars and clusters, has an integrated spectral classification of A5, and seems to contain a great deal of dust.

## 13.7 THE MASSES OF THE GALAXIES

As we know from our work with the planets of the solar system and the components of binary stars, the mass of a celestial body is determined from its gravitational effect on another celestial body. How, then, can we determine the masses of the galaxies, isolated as they are from one another in space? Let us consider the spiral systems. Their symmetry and form indicate that they are rotating about the central nuclei. Their enormous distances and sizes preclude any direct observation of this rotation within many hundreds of human lifetimes. However, when they are nearly, but not exactly, on edge as seen from the earth, the velocity of rotation at various distances from the central nuclei may be determined from the Doppler shift in the spectral lines of the stars. A star revolving about the center of a galaxy moves, because of galactic rotation, as if it were revolving about a mass concentrated at the center and equal to the mass of material inside its own orbit of revolution. When the linear velocity of approach or recession of a star well toward the edge of a galaxy that is inclined to the line of sight has been measured and the distance to the galaxy and consequently the distance of the star from its center are known, an estimate may be made of the mass. When the angular distance from the center of the galaxy and the velocity of rotation at that distance are known, finding the period of rotation in sidereal years is a matter of simple arithmetic. The distance in AU's can also be easily calculated. Once the period of revolution of a body and the distance from the center in appropriate units is known, we can use Kepler's law to calculate the mass in terms of the mass of the sun. The masses of extragalactic systems are provisionally estimated to range from below 10 million to a trillion or more solar masses. Since estimates of the mass of our own galaxy range from 200 to 400 billion solar masses, these conclusions appear to be reasonably consistent. However, the probable errors of these estimates are necessarily large, and the figure given for any particular galaxy should be considered only a first approximation.

The mass-luminosity relationship in stars tells us that those with a greater mass are also more luminous. This is not always true of galactic systems. The ratio between the mass and the luminosity of a galaxy is expressed in solar units. For example, if a galaxy has a mass of 10 billion solar masses and a total integrated luminosity 10 billion times the luminosity of the sun, its mass-lumonosity ratio is 1. If its mass was 10 times as great and its luminosity was unchanged, its mass-lamentity ratio would be 10. Fundamentally, this ratio seems to range from 50 to 1. When the mass-luminosity ratio is higher than, say, 30, this indicates that the galaxy consists mostly of stars of low luminosity (fainter than the sun and on the lower part of the main sequence). When the mass-luminosity ratio is low (that is, the proportionate luminosity is high), stars of high luminosity (on the upper part of the main sequence) probably predominate.

## 13.8 CLUSTERS OF GALAXIES

The Milky Way is a member of what is called the *Local Group* of galaxies. This is a cluster of about a dozen and a half known galaxies occupying an ellipsoidal volume of space about 1 mpc long. The Milky Way is near one end of the group and M31 in Adromeda is near the other. Of the 17 certain members of this group, 3 are spiral, 4 are irregular, and 10 are elliptical. Some of the elliptical systems are quite faint. There may be other members of the Local Group and some suspected members have been suggested. The obscuring material in the Milky Way could very easily hide a number of galaxies, and many others might be too faint to

Figure 13.15 **The Coma Cluster**
This photograph shows part of the regular cluster of galaxies in Coma Berenices.
Photograph from the Hale Observatories.

be recognized for what they are as galaxies. The combined mass of the Local Group is estimated to exceed 500 billion solar masses.

There are other small groups of galaxies similar to the Local Group near us. Their distances are fairly well established, since they are close enough for Cepheid variables to be observed. These groups, however, are representative only of small clusters.

The large groups of galaxies have been divided into two classes: regular and irregular clusters. Regular clusters resemble globular clusters of stars and are sometimes referred to as globular clusters. They have, as one might expect, a high degree of spherical symmetry and central condensation. They are rich in galaxies, but consist pri-

Figure 13.16 **The Hydra Cluster**
This is a distant cluster of galaxies, perhaps as far as 500 mpc from us.
Photograph from the Hale Observatories.

marily of elliptical systems and S0 spirals, the other types of spirals being absent or very rare. The nearest is in the constellation of Coma (Figure 13.15) at a distance of about 100 mpc. Over 2000 members of this rich cluster have been counted.

*Irregular*, or *open*, *clusters* do not have the spherical symmetry or central condensation of the regular clusters. They have many subgroups and contain all types of galaxies. They are more numerous than globular clusters. The nearest is in Virgo, 10 to 20 mpc away, and contains about a thousand galaxies. A more distant cluster in Hydra is shown in Figure 13.16.

An attempt has been made, with some degree of success, to detect evidence of *superclusters*, or arrangements of clusters, in the visible universe, but results in this area are still extremely tentative.

381

About 1912, it had been noticed that the exterior systems (which at that time were not recognized as other galaxies) had large Doppler shifts, indicating a high velocity. Of course, some of this velocity was due to the observer's motion, produced by the then-unrecognized rotation of our own galaxy. The velocities measured seemed to be mostly velocities of recession, although a few were velocities of approach.

At this early date, observations were confined to galaxies within or near 20 mpc. However, by 1930 enough velocities had been observed to permit some generalizing. The Doppler principle tells that the velocity, $v$, of approach or recession of an object is related very simply to the shift in wavelength of a spectral line. In fact, if $\lambda_0$ is the wavelength of a particular spectral line as measured in the laboratory and $\lambda$ is the wavelength of this same line as measured in the spectrum of a moving object such as a star or galaxy, then

$$v = c\left(\frac{\lambda - \lambda_0}{\lambda_0}\right)$$

where $c$ is the speed of light. If, for convenience, we define

$$\frac{\lambda - \lambda_0}{\lambda_0}$$

as equal to $z$, the Doppler equation then becomes

$$v = cz$$

These formulas allow us to determine the velocity of approach or recession of any source of electromagnetic radiation from an observation of the shift in the lines of its spectrum.

When Hubble had accumulated two sets of information (velocities and distances) for a sizable number of galaxies, he was able, in 1930, to announce that the velocity of recession of the galaxies (none had velocities of approach) was equal to a constant times their distance. The mathematical expression of this statement, where

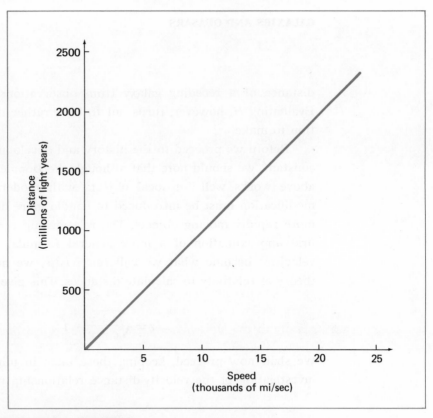

**Figure 13.17  The Velocity-Distance Relationship**
Measuring the red shifts of distant galaxies tells us how fast they are moving away from us. When the recessional velocities of galaxies and their distances are graphed, the data fall along a straight line whose slope equals the Hubble constant.

$v$ is the velocity of recession, $H$ is a constant, and $r$ is the distance of the galaxy, takes the following form:

$$v = Hr$$

Hubble's data on a variety of galaxies permitted him to evaluate the constant $H$. Their observed red shifts had given him a set of velocities, and observations of Cepheids, bright stars, integrated luminosities, and novas and supernovas had given him a set of distances for those objects. If his deductions were correct, graphing the distance and velocity of these objects should produce a straight line whose slope would be the constant $H$ (Figure 13.17). Once this constant is evaluated, it should then be possible to compute the distance of a receding galaxy from observations of its velocity. Evaluating $H$, however, turns out to be a rather difficult computation to make.

Before we proceed to the history and development of Hubble's constant, we should note that although the Doppler equation given above works well for local objects with moderate velocities, a modification must be introduced to handle more distant and hence more rapidly moving objects. The relationship $v = cz$ is simply a first approximation of a more general formula. Moreover, when velocities become what we call *relativistic*, we must turn to the theory of relativity to calculate distances. This gives us the formula

$$z = \sqrt{\frac{c + v}{c - v}} - 1$$

We shall now proceed, keeping these facts in mind, to a further investigation of the velocity-distance relationship $v = Hr$.

## 13.10 THE HUBBLE CONSTANT

In all measures of radial velocities of stars, a reduction is made from the observed velocity to the velocity relative to the sun by removing the velocity introduced by the orbital motion of the earth. In a study of galaxies, it is necessary to remove similar spurious components of velocity. One of these spurious components is that introduced by galactic rotation. The sun, and hence the solar system, is moving about the center of the galaxy at approximately 150 miles per second. This motion must be considered in all measurements of extragalactic velocities. Another movement that must be considered in measuring moderate and distant galactic velocities is the movement of our own galaxy in the Local Group. This is determined in much the same way as solar motion, that is, from observations of nearby objects with reference to which our velocity is measured. Finally, in measuring very distant objects, it is necessary to consider the movement of the Local Group with respect to the cluster of galaxies of which it is a part.

The independent determination of Hubble's constant depends on an equally independent determination of the distances of known galaxies. Astronomer Allan Sandage's limit for reliable independent distance estimates is from 3 to 20 mpc. Closer than 3 mpc, individual spurious velocities dominate and distort the estimates. Twenty mpc is the upper limit for reliable observations of distance indicators (such as bright stars) with the 200-inch telescope. Galaxies within these limits, then, are used to determine Hubble's constant.

For many years, the generally accepted value of $H$ was about 330 mi/sec/mpc. Thus once a figure for the velocity of an extra-galactic object was obtained, this value was used to compute its distance. A velocity of recession of 3300 miles per second meant that the galaxy was 10 mpc distant. A velocity of recession of 33,000 miles per second meant that the object was 100 mpc away. However, a careful consideration of the dimensions of $H$ reveals that $1/H$ is a measure of time. If the universe started from a very compact state and has been expanding at a uniform rate ever since, then $1/H$ must be the upper limit of the age of the universe. Considered in this light, the value for $H$ accepted between 1936 and 1952 would have to be wrong, since it gave an age for the universe of 1.8 billion years, less than the accepted age of the earth. If the value of $H$ was not wrong, then there was something the matter with the current model of the universe, which assumed a uniform expansion. Sir Arthur Eddington and Abbé Georges Lemaître proposed new models of the universe, using a repulsive force that increased the velocity to balance or overcome the deceleration. However, there was no evidence of this repulsive force; neither the theory of relativity nor Newtonian physics automatically provided for a repulsive force between particles that increased with the distance between them or varied in any other way. The other option, that $H$ was incorrect, was not considered seriously until

1952. At that time, Walter Baade showed that the period-luminosity law for the Cepheids was in error by 1.5 magnitudes. His correction increased the extragalactic distance scale by 2, effectively doubling the age of the universe, and cutting $H$ in half. Eventually, the period-luminosity curve was recalibrated and the distance of the members of the Local Group was recomputed. In 1968, the distance to the M81 cluster was multiplied by about 4.5, placing it 3.4 mpc away. Each recalibration of the period-luminosity curve has resulted in a decrease in the Hubble constant, and today the accepted value is approximately 30 mi/sec/mpc. This is roughly equal to 10 mi/sec per million light years. Thus the age of the universe seems to be about 20 billion years.

## 13.11 THE EXPANDING UNIVERSE

The full implications of Hubble's work with respect to the motion of the universe were not immediately accepted by all astronomers, and alternative explanations of the apparent Doppler shift were offered. It was suggested that in the millions and billions of years during which light traveled through space it somehow lost energy, thus causing a shift of spectral lines to the red or longer wavelengths. However, there was no foundation for this supposition, and the "tired light" theory, as it came to be known, was abandoned. Another explanation was based on the relativistic shift in the lines of a spectrum when the light leaves a strong gravitational field. This phenomenon is well known and has actually been observed in white dwarfs. However, the mass necessary in the galaxy to produce the observed shifts is much too great to be even very slightly probable. In the absence of any satisfactory alternative, then, astronomers assume that the universe is expanding. Since we are not certain whether our present value for the Hubble constant is correct, we cannot be certain what kind of expansion we are observing.

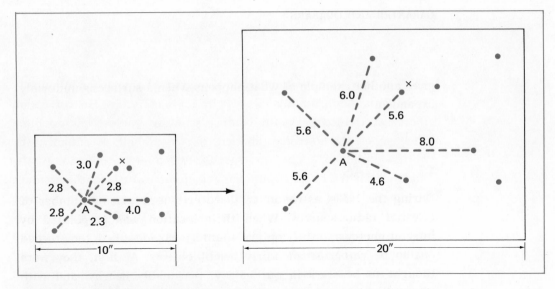

Figure 13.18 **The Expanding Universe**
In these two figures, a drawing is expanded from 10 inches on a side to 20 inches on a side. The distances from dot A to other dots in the expanded drawing are now twice what they were before. An observer at A feels that he is in the center of the expansion. However, if the position of any dot other than A is examined in the same way, it will be found that all the dots have moved correspondingly far away from it.

At first glance, it may appear that we are at the center of an expanding universe. Everything in creation is rushing away from us with ever-increasing velocity as its distance increases. The reason the distance and velocities increase together is simple. The fastest-moving galaxies are the farthest away. But how about our apparent position at the center? Actually, if one considers the matter carefully, it becomes evident that we are not necessarily at the center of the expansion, nor do we know where the center is. No matter what galaxy one picked as the position of the observer, all the other galaxies would be rushing away from it once spurious velocities were removed. Imagine a small balloon with many spots painted on it. If the balloon is inflated to twice its original diameter, every spot will be twice as far from every other spot as it was in the beginning. No matter which spot one observes this process from, the universe (which consists in this case of the surface of the balloon) has expanded and that particular spot appears to be the center of the expansion. The illustration above (Figure 13.18)

387

gives another example of what happens when a surface is uniformly expanded.

## 13.12 QUASARS

During the 1960s, astronomers discovered a surprising number of celestial radio sources. When their location was pinpointed by interferometry and their optical counterparts identified, they turned out to be rather small, fairly bright objects. At first, they were thought to be stars in our galaxy, reasonably nearby, and with the ability to produce a great deal of radio noise. However, the only true star from which a measurable radio signal had definitely been received was the sun. If similar radio signals were produced even by the sun's nearest neighbor, they would be undetectable at the distance of the earth. Moreover, the spectra of the radio sources showed some very strange lines, quite unlike those of the normal stellar spectra. Since they could not definitely be classified as stars, they were given the name quasi-stellar radio sources, which was soon shortened to *quasars*.

The first key to the puzzle of the quasars was provided by Martin Schmidt in the early 1960s. Schmidt was studying the spectra of the radio source 3C273 (Figure 13.19) when he noticed that four of the lines in the optical spectrum formed a simple harmonic pattern of separation and intensity decreasing toward the ultraviolet. This is characteristic of a hydrogen atom, or any other atom that has been stripped of all its electrons but one. Schmidt concluded that no atom had produced radiation of the observed wavelengths. He found that, if he assumed that the lines in the spectrum had been shifted toward the red by 16 percent, four of them looked like hydrogen lines. One of the remaining lines seemed to be doubly ionized oxygen, and the last line, near the ultraviolet end of the optical spectrum at 3229 Å, was an ionized magnesium line that had been shifted from 2800 Å in the

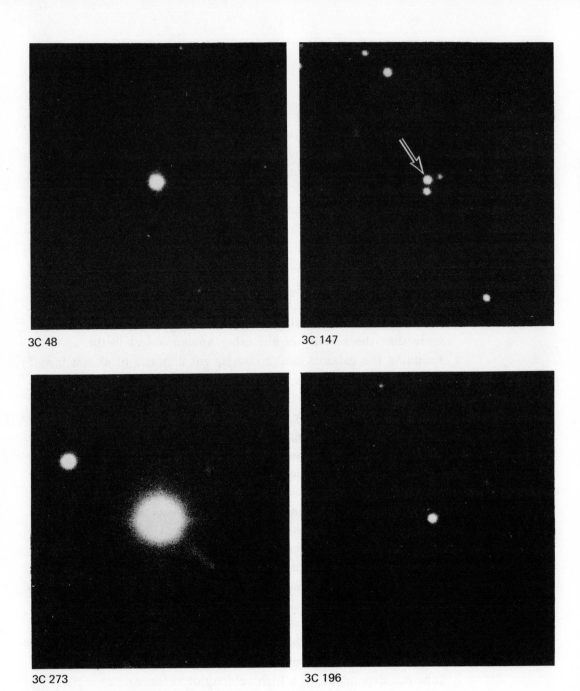

3C 48

3C 147

3C 273

3C 196

Figure 13.19 **Four Quasars**
Four of the most well-known quasars are shown in this photograph. Although
quasars look like very blue stars, their spectra show enormous red shifts.
Photograph from the Hale Observatories.

part of the ultraviolet spectrum that is normally unobservable because of the opacity of the earth's atmosphere. It was not the lines in the quasar's spectrum that were strange, but their position, which was due to an apparently tremendous red shift. Familiar lines were subsequently identified in the spectra of other quasars, and red shifts of up to 37 percent were observed. Figure 13.20 shows the quasar 3C295, which has one of the largest red shifts ever measured.

Such phenomenal red shifts implied a tremendous velocity of recession on the part of the quasars that would make them enormously far away. Some quasars had recessional velocities, according to their red shifts, of 80 percent of the velocity of light, more than the velocity of any other known object in the universe, including the galaxies, and a consequent distance of as much as 5 billion light years. How was it, then, that they were so bright as to be mistaken for nearby stars? Distant objects, unless they are intrinsically enormously luminous, always appear very faint. The quasars had visual magnitudes of about +13 to +19. There are galaxies with an apparent magnitude of about +13, but they are comparatively close and contain reasonably well-resolved stellar systems. The quasars simply do not look like galaxies. Moreover, they sometimes vary in brightness in a way that implies some sort of physical change. For this variation to have an observable period, they must be small enough for whatever agency causes it to work its way through their geometric structure in a reasonably short time. This would make them considerably smaller than even a moderately sized star cluster, let alone a galaxy. Their variation does not resemble that of a pulsar.

Various attempts have been made to solve the mystery of the quasars by suggesting a different explanation of their red shift (one not dependent on a high velocity of recession) or a different explanation of their velocity (one not dependent on participation in the expansion of the universe for a long enough time to make them impossibly distant or impossibly bright). None have been

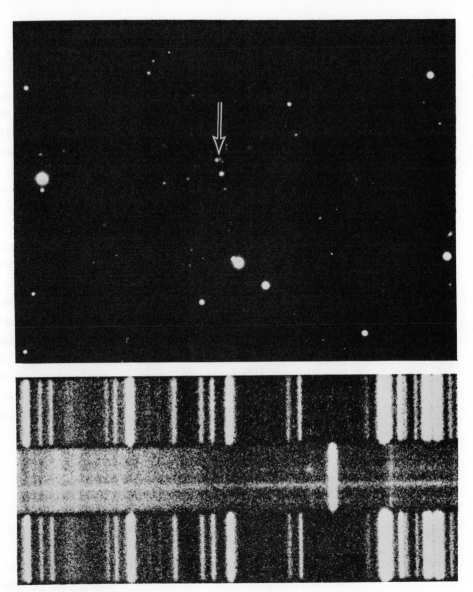

Figure 13.20 **The Quasar 3C295 in Bootes**
This quasar has one of the largest red shifts ever measured. Photograph from
the Hale Observatories.

particularly successful, however. Red shifts can be caused by
relativistic gravitational fields, but for the quasars to have such
fields they would have to be impossibly dense and would collapse
into themselves, forming black holes. It is conceivable that the

quasars are nearby enough to account for their apparent brightness if we assume that they were latecomers to the expansion of the universe and hence closer than their recessional velocity indicates. However, the spectra of some quasars contain absorption lines associated with the Virgo cluster of galaxies, which is about 12 mpc away. For radiation from the quasars to be passing through this cluster, they must, obviously, be farther away. It has also been suggested that the quasars may be objects ejected from galaxies; but if this is so then some of them, at least, should be moving toward us and have observable blue shifts. None do.

The mystery of the quasars, then, centers on the red shifts in their spectra. If these shifts are due to their velocities of recession, then they are enormously far away and enormously luminous, emitting thousands of times more energy than ordinary galaxies. If they are not enormously far away, their red shifts are inexplicable. Whatever the eventual explanation of the quasars may be, it is likely to have a far-reaching effect on our concept of the nature, behavior, and formation of the physical universe.

# QUESTIONS

(1) What was the Shapley-Curtis debate about? How did Hubble settle it? (Section 13.1)

(2) How are Cepheid variables used in determining the distances to nearby galaxies? (Section 13.1)

(3) How are the distances to more distant galaxies determined? (Section 13.2)

(4) What discovery led to the idea that the universe is expanding? (Section 13.2)

(5) Describe the four basic types of galaxies. (Section 13.3)

(6) Which are more common: spiral galaxies, or barred spiral galaxies? How do they differ? (Section 13.4)

(7) Why are elliptical galaxies thought to be the oldest galaxies? (Section 13.5)

(8) How do we infer the masses of galaxies? (Section 13.7)

(9) What is the Local Group? (Section 13.8)

(10) Explain the velocity-distance relationship. (Section 13.9)

(11) What is the significance of the Hubble constant? (Section 13.10)

(12) If all the galaxies are observed to be moving away from us, why can we not conclude that we are at the center of the universe? (Section 13.11)

(13) Why are quasars among the most puzzling celestial phenomena? (Section 13.12)

# CHAPTER 14
# COSMOLOGY

Our final task in our study of the universe is to synthesize our knowledge and formulate
a general theory from what we have learned. The term *cosmology* has come to mean the
study of the origin, subsequent evolution, present organization, and probable future of
the universe. The laws governing this study are indispensable to our understanding of
what we see in space. At the very foundation of cosmology is the basic question of what
type of geometry best describes the universe. The answer both underlies and transcends
all the other answers found in the long history of man's efforts to solve the cosmological
problem.

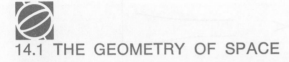

## 14.1 THE GEOMETRY OF SPACE

Any efforts to understand the universe require certain basic assumptions. Hence we assume that when we are dealing with ordinary distances and velocities, the laws of Newtonian physics are a valid guide to the interpretation of what we observe. The behavior of satellites, planets, double stars, and even our galaxy seems to affirm this. However, there are instances in which the theory of relativity is a necessary tool. Such phenomena as the advance of Mercury's perihelion (Section 6.1), the bending of light rays passing the sun (Section 4.13), the escape of electromagnetic radiation from intense gravitational fields (Section 11.7), and the phenomenon of collapsars (Section 11.9), for example, can be explained only by turning to Einstein's model of the universe.

The fundamental concept behind Einstein's theory of relativity is that a gravitational field affects the geometry of space and time. The conclusions to which this assumption leads seem to be incompatible with ordinary experience. This is due, at least in part, to the geometric assumptions we normally make. *Geometry* means, literally, "to measure the earth," but its modern definition is the study of abstract structure and form. Any cosmological explanation of the universe must have a geometric foundation. In our daily lives, as well as in most scientific studies, we take for granted the ancient concepts of *Euclidean* geometry. In this geometry, the sum of the angles of triangles is 180°, parallel lines never meet (or meet at infinity), and the area of circles is given by the formula $\pi r^2$, where $r$ is the radius. Many other everyday, or "commonsense," conclusions follow from what we call logical reasoning.

In the nineteenth century, a suspicion gradually arose in the scientific community that, first, Euclidean geometry was not the only possible geometry and, second, that an alternative geometry might provide a better explanation of the cosmos. From these suspicions two non-Euclidean geometries emerged: the *hyperbolic* geometry of Lobachevski and the *elliptical* of Riemann.

All geometries are constructed from basic assumptions called *axioms*. Although we need not examine the axioms of the various

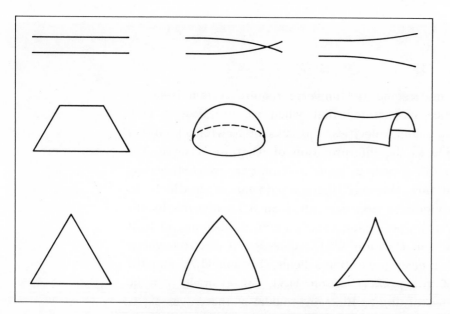

Figure 14.1 **Three Possible Geometries**
What happens to parallel lines and angles tells us about the
curvature of space. There are three possibilities. Either space is *flat*
(has zero curvature), in which case parallel lines never meet and the sum
of the angles of a triangle (called the *angle sum*) is 180°;
or space is *positively curved*, in which case parallel lines are
impossible and triangles have angle sums greater than 180°; or space
is *negatively curved*, in which case more than one parallel to a given line
may be drawn through a point not on that line and triangles
have angle sums less than 180°.

geometries here, the conclusions they produce (Figure 14.1) are
interesting. In hyperbolic geometry, the sum of the angles of a
triangle is less than 180° (the exact sum depends on the size of
the triangle) and more than one parallel to a given line may be
drawn through a point not on that line. In elliptical geometry, the
sum of the angles of a triangle is more than 180° (again, the exact
sum is a variable related to the triangle's size) and parallel lines
do not exist.

Models of various geometries may be constructed. Euclidean
geometry is demonstrated on a plane (uncurved) surface. Elliptical
geometry is demonstrated on the surface of a sphere (a surface
with *positive* curvature). For example, the sum of the angles of
the triangle on the earth formed by the equator and two meridians
of longitude, 0° and 90°, is 270°. Any two great circles drawn on the

surface of a sphere will meet in two opposite points; thus parallels are impossible.

Hyperbolic geometry may be represented on a saddle-shaped surface (a surface with *negative* curvature). The sum of the angles of triangles drawn on such surfaces is always less than 180°; the other properties of hyperbolic geometry are also evident.

It is very difficult to visualize the results of geometries other than Euclidean geometry in space of more than two dimensions, or the results of *any* geometry in space of more than three dimensions. Yet it is clear that we are not restricted to Euclidean geometry in our explanation of the cosmos. Observational tests do exist that can, theoretically at least, answer the fundamental question of which geometry best represents the true geometry of the cosmos. We shall consider these tests in the discussions that follow.

## 14.2 DEVELOPING A PORTRAIT OF THE UNIVERSE

In many ways, cosmology parallels the study of the evolution of the stars, but there are differences. First, the situation is immeasurably complicated by the sheer magnitude of the universe. A star is a discrete body. It is a member of a large population of discrete bodies, many similar and many different. A study of the universe, however, is a study of a unique entity. There is, by definition, no other. We may study different parts of the cosmos, but there is no other complete universe with which we can compare our own. If we assume, as many have, that the galaxies, the basic components of the universe, are all the same age, we are again at a disadvantage. In our study of the evolution of the stars, although we cannot witness the entire life cycle of a single individual, we have spread out before us a panorama of stars of different ages—a panorama, as it were, of stellar evolution. We do, however, have what is to some extent a compensating advantage

Figure 14.2 **A Cluster of Galaxies in Corona Borealis**
This group of galaxies is probably over 200 mpc away. Photograph from the Hale Observatories.

in our study of galaxies. When we look at galaxies many mega-parsecs away, like those in Figure 14.2 and Figure 14.3, we are also looking back into time. The light that reaches us from them began its journey through space ages ago; hence it gives us a clue as to how the galaxies appear in the past.

Every civilization has developed a cosmology. The Hebrews, for example, recorded theirs in the *Book of Genesis*. Each cosmology depends on the knowledge and sophistication of the society that develops it. Thus our twentieth-century cosmology exhibits what we like to think of as the sophistication of our time. In the future it may be regarded as an extremely primitive effort. How-ever, our concept of the cosmos is an expression of what we know

398

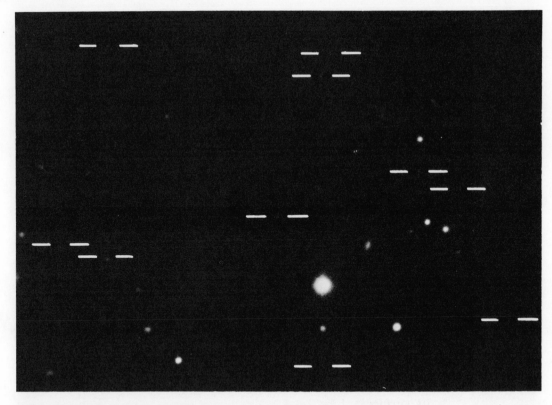

Figure 14.3 **Faint Galaxies in Coma**
This cluster (which is not the same as that in Figure 13.15) is also very distant.
Photograph from the Hale Observatories.

and what we have theorized. It is derived from our knowledge of the celestial universe and the laws of modern physics.

Before we proceed further, we should discuss what is known as the *cosmological principle*. In Chapter 10 we mentioned that astronomers in the late 1920s found the earth to be peculiarly located at the center of a system of clusters. Moreover, the farther away from the earth these galactic clusters were, the larger their linear diameter seemed to be. Whenever an observer finds himself in a peculiar position that does not have an immediate logical explanation, he should, according to the cosmological principle, investigate why this anomaly exists. Briefly stated, this principle is,

> Local variations aside, all observers everywhere should
> see the universe in essentially the same form.

This principle introduces a number of interesting problems. For example, if it is strictly true, there can certainly be no edge to the universe, since an observer on the edge would see all manner of celestial objects on one side of the sky and empty space on the other while an observer at the center would see material in all directions. There are various ways of overcoming this particular problem. One is to assume that the universe is infinite. Another is to assume that the basic geometry of the cosmos is four-dimensional and has a nonzero curvature. Elliptical geometry can be represented by the two-dimensional curved surface of a three-dimensional sphere. The postulation of a curved four-dimensional space-time continuum helps explain some of our observations of the universe. A space that curves back on itself is an analogy of four-dimensional space in somewhat the same way that the surface of a sphere is a two-dimensional analogy in a three-dimensional space. No one would argue that the surface of the earth is infinite in extent. It is obvious that there is no termination or edge to its surface.

## 14.3 MODELS OF THE CELESTIAL UNIVERSE

The discovery in the early part of the twentieth century that the galaxies were receding, and the universe was therefore expanding at an enormous rate, has given rise to various theories of the origin and evolution of the cosmos. We shall examine what seem to be the most likely explanations below.

## 14.4 THE BIG BANG

One way to explain the expansion of the present universe indicated by the red shifts of the galaxies is to postulate the explosion of a denser universe far back in time. This is known as the *big-bang theory* of cosmic evolution. According to this theory, about 10 or 20 billion years ago the universe existed in a highly concentrated state. Something caused this "primeval atom" to explode, and the material that later condensed into galaxies was impelled outward in all directions. The fastest-moving galaxies are now the most distant from the original explosion. We cannot locate the center of the expansion, the point at which the explosion occurred, any more than a miniature observer standing on the expanding balloon described in Chapter 13 could find a "central" dot.

We noted in Chapter 13 that if the expansion of the universe were uniform, the reciprocal of the Hubble constant, $1/H$, would be the upper limit of the age of the universe. Actually, it will not last quite this long, since the gravitational attraction of everything in the universe for every other thing slows the expansion down over time. Just how much it slows the expansion down turns out to be a rather crucial question. If the constant of deceleration, $q_0$, is greater than zero but less than $+0.5$, then the geometry of the universe is hyperbolic (open) and the expansion will continue indefinitely. It is impossible, at present, to evaluate the constant of deceleration with any accuracy. However, Dr. William A. Baum's recent work with the 200-inch telescope has produced a tentative estimate of $+0.3$. This, then, would support the theory of a hyperbolic, ever-expanding universe. On the other hand, if $q_0$ is greater than $+0.5$, then the geometry of the universe is elliptical (closed) and the expansion will eventually come to a halt. This brings us to our next theory of cosmic evolution, that of a pulsating universe.

## 14.5 THE PULSATING UNIVERSE

If the constant of deceleration, which depends ultimately on the total amount of mass in the universe, is large enough to eventually halt the observed expansion, an era of contraction may start. This could result in a pulsating, or oscillating, universe that alternately

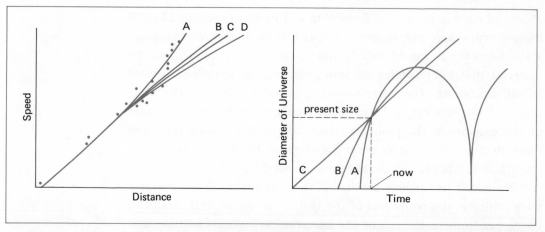

Figure 14.4 **Four Cosmological Models**
A plotting the distance and speed of recession of the most distant galaxies may
not produce the straight line postulated by Hubble. In the diagram on the left,
curves A, B, and C are for evolving (big-bang) universes with different values of
$q^0$, the deceleration parameter. Curve D is for a steady-state universe. Curve A is
for a positively curved, pulsating universe. Curves B and C are for open
universes having flat and hyperbolically curved space, respectively. As the diagram
on the right shows, once we know which curve is correct, we can make a
definite statement about the past and future size of the Universe.

expands and contracts. The reciprocal of the Hubble constant, adjusted for the constant of deceleration, would then be the upper
limit, not of the age of the universe, but of the present era of
expansion. The universe would have no ascertainable beginning or
end.

Allan Sandage has given a tentative estimate of $q_0$ of $1.2 \pm 0.4$,
but warns against taking it very seriously. However, if this value
of $q_0$ is combined with the estimated value of the reciprocal of
$H$ (which is about 10 billion years), it gives an estimate of the
beginning of the present expansion of from 4.2 to 11.7 billion years
ago. As Sandage points out, these figures are compatible with estimates of the age of the oldest stars and the chemical elements
(about 10 billion years). Figure 14.4 shows the effect of changes in
$q_0$ on the speed of recession of the galaxies and the size of the
universe.

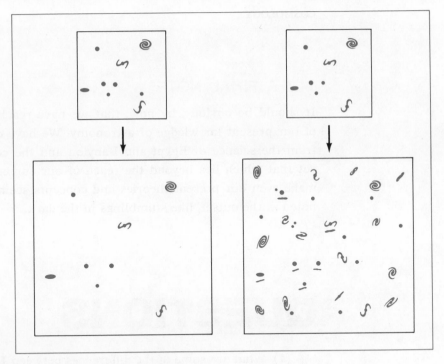

**Figure 14.5 The Steady-State Theory vs. the Big-Bang Theory**
The differences between these two cosmologies can be illustrated by examining
what happens to a volume of space over a long period of time. If the big-bang
theory is correct, then the galaxies simply get further apart, as in the diagram on
the left. In the steady state theory, as the older galaxies get further apart, new
galaxies are created to fill the resulting void, as shown in the diagram on the right.

## 14.6 THE STEADY-STATE UNIVERSE

The theory of a steady-state universe, one in which material is
constantly being created, so that the density remains the same
despite the recession of existing galaxies, is in a sense an extension
of the cosmological principle to mean that all observers everywhere
and *at all times* should see the universe in essentially the same
way. It was a rather prominent theory some decades ago, but
whether it should now be taken seriously is open to question.
Such an explanation of the cosmos would require a negative constant of deceleration, and the evidence now seems to point away
from this. Figure 14.5 shows the type of universe that would
result from the steady-state and the big-bang theories.

## 14.7 A FINAL NOTE

It should be obvious, by now, that we have reached the frontiers of our present knowledge of astronomy. We have come a long way from the science of Egypt and Babylon and the court astrologers, yet that which lies beyond the reach of our current probings may make even our present theories and concepts seem one day, as we noted at the outset, like stumblings in the dark.

# QUESTIONS

(1) What are some of the differences between Euclidean, hyperbolic, and elliptical geometry?                    (Section 14.1)

(2) How can we, in our study of the galaxies, "look back" in time?                    (Section 14.2)

(3) What is the cosmological principle?          (Section 14.2)

(4) Why do we expect the rate of expansion of the universe to slow down?                    (Section 14.4)

(5) What evidence is there for a pulsating universe? for an ever-expanding universe?          (Sections 14.5 and 14.6)

# GLOSSARY
# APPENDIXES
# INDEX

# GLOSSARY

**absolute magnitude** A measure of the real brightness of celestial objects. Technically, the apparent magnitude a star would have at a distance of 10 parsecs.

**absolute zero** The lowest possible temperature; a temperature of 0°K or −273°C.

**absorption spectrum** Dark lines superimposed on a continuous spectrum of electromagnetic radiation.

**acceleration** A change in velocity; speeding up, slowing down, or changing direction.

**active sun** The sun during times of unusual activity; a period of sunspots, flares, and associated phenomena.

**albedo** The percentage of sunlight reflected by a planet, a satellite, or an asteroid.

**almanac** A book or table listing astronomical events.

**altitude** Angular distance above or below the horizon, measured along a vertical circle, to a celestial object.

**amplitude** A range of variability, as in the light from a variable star.

**angstrom (Å)** A unit of length equal to one hundred-millionth of a centimeter.

**angular size** The angle subtended by an object.

**annular eclipse** An eclipse of the sun in which the moon is too distant to cover the sun completely, so that a ring of sunlight shows around the moon.

**aphelion** The point in a planet's orbit at which it is farthest from the sun.

**apogee** The point in the orbit of a satellite at which it is farthest from its primary.

**apparent magnitude** A measure of the brightness of a star or another celestial object viewed from the earth.

**apparent relative orbit** The projection onto a plane perpendicular to the line of sight of the relative orbit of the fainter of the two components of a visual binary star about the brighter.

**apparent solar day** The interval between two successive transits of the sun's center across the meridian.

**apparent solar time** Time reckoned by the actual position of the sun in the sky.

**ascending node** The point in the orbit of a celestial body where it crosses from the south to the north

of a reference plane, usually the plane of the celestial equator or the ecliptic.

**aspect** The position of the sun, satellites, or planets with respect to one another.

**association** A loose cluster of stars whose motions, spectral types, or positions in the sky indicate a common origin.

**asteroid** A minor planet.

**astrometric binary** A binary star in which one component is not observable, but can be deduced from the motions of the visible star.

**astronautics** The science of laws and methods of spaceflight.

**astronomical unit (AU)** The average distance between the earth and the sun; a unit of length equal to approximately 93 million miles.

**astronomy** The study of the universe beyond the earth's atmosphere.

**astrophysics** The study of the physical properties and phenomena associated with planets, stars, and galaxies.

**atom** The smallest particle of an element that retains the properties which characterize that element.

**aurora** Light radiated by atoms and ions high in the earth's atmosphere, sometimes called "the northern lights."

**autumnal equinox** An intersection of the ecliptic and the celestial equator; the point at which the sun crosses the celestial equator moving from north to south.

**azimuth** The angle along the celestial horizon, measured eastward from the north point, to the intersection of the horizon with the vertical circle passing through an object.

**barred spiral** A spiral galaxy in which the arms project from the ends of a bar running through the nucleus.

**baseline** The side of a triangle whose length is known and which is used to determine the position of an object at the apex.

**big-bang theory** A cosmological theory in which the expansion of the universe is presumed to have begun with a primeval explosion.

**binary star** A double star; two stars revolving about each other.

**blackbody** A hypothetical, perfect radiator; a body that absorbs and reemits all the radiation falling on it.

**blink microscope** A microscope in which the user's view is shifted rapidly back and forth between two different photographs of the same region of the sky.

**Bode's law** A technique for obtaining a sequence of numbers giving the approximate distances of the planets from the sun in astronomical units.

**Bohr atom** A model of the atom developed by Niels Bohr in which the electrons revolve about the nucleus in circular orbits.

**bolide** A very bright, or fireball, meteor; a fireball accompanied by sound.

**bolometric correction** The difference between the visual and bolometric magnitude of a star.

**bolometric magnitude** A measure of the total flux of radiation from a celestial object received just outside the earth's atmosphere and detected by a device sensitive to *all* forms of electromagnetic energy.

**celestial equator** A great circle on the celestial sphere, 90° from the celestial poles.

**celestial mechanics** A branch of astronomy that deals with the motions of the members of the solar system.

**celestial navigation** The art of navigating at sea or in the air from sightings of the sun, moon, planets, and stars.

**celestial poles** The points about which the celestial sphere appears to rotate; the points of intersections of the celestial sphere with the extension of the earth's polar axis.

**celestial sphere** The apparent sphere of the sky; a sphere of large radius centered on the earth.

**Cepheid variable** A star that belongs to one of two classes (type I and type II) of yellow, supergiant, pulsating stars.

**Ceres** The largest of the minor planets and the first to be discovered.

**chromatic aberration** A defect in an optical system that causes light of different colors to be focused at different points.

**chromosphere** The part of the solar atmosphere immediately above the photosphere.

**circumpolar region** Portions of the celestial sphere, near the celestial poles, that are either always above or always below the horizon.

**cluster of galaxies** A system of galaxies with at least several and in some cases thousands of members.

**cluster variable** A member of a certain class of pulsating variable stars, all with periods less than 1 day; an RR Lyrae variable.

**color excess** The amount by which the color index of a star increases when its light is reddened in passing through interstellar absorbing material.

**color index** The difference between the magnitudes of a star measured in light from two different regions of the spectrum.

**color-magnitude diagram** A plot of the apparent or absolute magnitudes of the stars in a cluster against their color indexes.

**comet** A swarm of solid particles and gases that revolves about the sun, usually in a highly eccentric orbit.

**comet head** The main part of a comet, consisting of a nucleus and a coma.

**cometary coma** The diffuse, gaseous component of the head of a comet.

**configuration** Any of several orientations in the sky of the moon or a planet with respect to the sun.

**conic section** The curve of intersection between a circular cone and a plane; an ellipse, circle, parabola, or hyperbola.

**conjunction** The configuration in which a planet appears nearest to the sun or some other planet.

**constellation** A configuration of stars named for a particular object, person, or animal.

**continuous spectrum** A spectrum

of light comprised of radiation of a continuous range of wavelengths, or colors.

**corona** The outer atmosphere of the sun.

**coronagraph** An instrument for photographing the chromosphere and corona of the sun when it is not eclipsed.

**cosmic rays** Atomic nuclei (mostly protons) that strike the earth's atmosphere at exceedingly high speeds.

**cosmogony** The study of the origin of the universe.

**cosmological model** A specific theory of the organization and evolution of the universe.

**cosmological principle** The assumption that, on the average, the universe is the same everywhere at any given time.

**cosmology** The study of the organization and evolution of the universe.

**crater (lunar)** A circular depression in the surface of the moon.

**crater (meteoritic)** A terrestrial crater caused by the collision of a meteoroid with the earth.

**crescent moon** A phase of the moon during which its elongation from the sun is less than 90° and it appears less than half full.

**dark nebula** A dark cloud of inter-stellar dust that obscures the light of more distant stars.

**daylight savings time** A time 1 hour ahead of standard time, usually adopted in spring and summer to take advantage of long evening twilights.

**declination** Angular distance north or south of the celestial equator.

**deferent** A stationary circle in the Ptolemaic system along which the center of another circle (an epicycle) moves.

**density** The ratio of the mass of an object to its volume.

**descending node** The point in the orbit of a celestial body where it crosses from the north to the south of a reference plane, usually the plane of the celestial equator or the ecliptic.

**differential gravitational force** The difference between the gravitational force exerted on two bodies near each other by a third, more distant body.

**differential rotation** Rotation in which all the parts of an object do not behave like a solid. The sun and galaxies exhibit differential rotation.

**diffraction grating** A system of closely spaced equidistant slits through which light is passed to produce a spectrum.

**411**

**diffuse nebula**  A reflection, or emission, nebula produced by interstellar matter.

**disk (galactic)**  The central, wheel-like portion of a spiral galaxy.

**dispersion**  The separation of white light into different colors, or wavelengths.

**distance modulus**  The difference between the apparent and absolute magnitude of an object.

**diurnal circle**  The apparent movement of a star during 24 hours due to the earth's rotation.

**diurnal libration**  A change in the visible hemisphere of the moon during the day because of the earth's rotation.

**diurnal motion**  Motion during one day.

**diurnal parallax**  An apparent change in the direction of an object caused by a displacement of the observer due to the earth's rotation.

**Doppler shift**  An apparent change in wavelength of the radiation from a given source due to its relative motion in the line of sight.

**dwarf (star)**  A main-sequence star with a low mass compared to that of a giant or supergiant star.

**eclipse**  The cutting off of all or part of the light of one body by another body passing in front of it.

**eclipse path**  The track along the earth's surface swept out by the tip of the shadow of the moon during a solar eclipse.

**eclipse season**  A period during the year when an eclipse of the sun or moon is possible.

**eclipsing binary star**  A binary star in which the plane of revolution of the two components is nearly edge on to our line of sight, so that one star periodically passes in front of the other.

**ecliptic**  The apparent annual path of the sun on the celestial sphere.

**ecliptic limit**  The maximum angular distance of the moon from a node at which an eclipse can take place.

**electromagnetic radiation**  Radiation consisting of waves propagated with the speed of light. Radio waves, infrared, visible, and ultraviolet light, x-rays, and gamma rays are all forms of electromagnetic radiation.

**electromagnetic spectrum**  The entire range of electromagnetic waves.

**electron**  A negatively charged subatomic particle that normally moves about the nucleus of an atom.

**ellipse**  A conic section; the curve of intersection of a circular cone and a plane cutting completely through it.

**elliptical galaxy** A galaxy whose apparent contours are ellipses and which contains no conspicuous interstellar material.

**elongation** The difference between the celestial longitudes of a planet and the sun.

**emission line** A discrete, bright spectral line.

**emission nebula** A gaseous nebula that derives its visible light from the fluorescence of ultraviolet light from a star.

**emission spectrum** A spectrum of emission lines.

**energy level** A particular level, or amount, of energy possessed by an atom or ion above that which it possesses in its ground state.

**ephemeris** A table giving the positions of a celestial body at various times or other astronomical data.

**ephemeris time** Time that passes at a strictly uniform rate, used to compute the instant of various astronomical events.

**epicycle** A circular orbit of a body in the Ptolemaic system, the center of which revolves about another circle (the deferent).

**equation of time** The difference between apparent and mean solar time.

**equator** A great circle on the earth, 90° from the North and South Poles.

**equatorial mount** A mounting for a telescope, one axis of which is parallel to the earth's axis, so that a motion of the telescope about that axis compensates for the earth's rotation.

**equinox** One of the intersections of the ecliptic and the celestial equator.

**eruptive variable** A variable star whose changes in light are erratic or explosive.

**evolutionary cosmology** A cosmology which assumes that all parts of the universe have a common age and evolve together.

**excitation** The imparting of energy to an atom or an ion.

**extinction** The attenuation of light from a celestial body produced by the earth's atmosphere or by interstellar absorption.

**extragalactic** Beyond the galaxy.

**eyepiece** A magnifying lens used to view the image produced by the objective lens of a telescope.

**faculus (pl. faculae)** Bright region near the limb of the sun.

**filar micrometer** A device used with a telescope to measure the angular separation of close pairs of stars or the diameters of extended objects.

**filtergram**  A photograph of the sun taken through a special narrow-bandpass filter.

**fireball**  A spectacular meteor.

**fission**  The breakup of a heavy atomic nucleus into two or more lighter ones.

**flare (solar)**  A sudden and temporary outburst of light from an extended region of the solar surface.

**flare star**  A member of a class of stars that show occasional, sudden, and unpredicted increases in light.

**flash spectrum**  The spectrum of the visible limb of the sun, obtained just before totality in a solar eclipse.

**flocculus (pl. flocculi)**  A bright region of the solar surface observed in the monochromatic light of a spectral line; usually called a *plage*.

**focal length**  The distance from a lens or mirror to the point where light converged by it is focused.

**focus**  The point where the rays of light converged by a mirror or lens meet.

**forbidden lines**  Spectral lines not usually observed under laboratory conditions because they result from highly improbable atomic transitions.

**Fraunhofer line**  An absorption line in the spectrum of the sun or a star.

**Fraunhofer spectrum**  The array of absorption lines in the spectrum of the sun or of a star.

**frequency**  The number of vibrations in a unit of time; the number of waves that cross a given point in a unit of time.

**full moon**  A phase of the moon during which it is in opposition (180° from the sun) and its full daylight hemisphere is visible from the earth.

**fusion**  The building up of heavier atomic nuclei from lighter ones.

**galactic cluster**  An open cluster of stars in the spiral arms or disk of the galaxy.

**galactic equator**  The intersection of the principal plane of the Milky Way with the celestial sphere.

**galactic latitude**  Angular distance north or south of the galactic equator measured along a great circle passing through the galactic poles.

**galactic longitude**  Angular distance measured eastward along the galactic equator from the galactic center to the intersection of the galactic equator with a great circle passing through the galactic poles.

**galactic poles**  The poles of the galactic equator; the intersections with the celestial sphere of a line perpendicular to the plane of the galactic equator.

**galactic rotation** The rotation of the galaxy.

**galaxy** A large assemblage of stars. A typical galaxy contains millions to hundreds of billions of stars.

**gamma rays** Photons of electromagnetic radiation with wavelengths shorter than those of x-rays. Gamma rays have the highest frequency of any form of electromagnetic radiation.

**giant (star)** A highly luminous star with a large radius.

**gibbous moon** A phase of the moon during which more than half, but not all, of the moon's sunlit hemisphere is visible from the earth.

**globular cluster** A large spherical cluster of stars.

**globule** A small, dense, dark nebula, possibly a protostar.

**granulation (solar)** The ricelike pattern evident in photographs of the solar photosphere.

**gravitation** The attraction of matter for matter.

**great circle** A circle on the surface of a sphere that is the curve of intersection of the sphere with a plane passing through its center.

**Greenwich meridian** The meridian of longitude passing through the site of the old Royal Greenwich Observatory, near London; the great circle from which terrestrial longitude is measured.

**Gregorian calendar** A calendar (now in common use) introduced by Pope Gregory XIII in 1582.

**H I region** A region of neutral hydrogen in interstellar space.

**H II region** A region of ionized hydrogen in interstellar space.

**Hertzsprung-Russell (H-R) diagram** A diagram showing the relationship of the absolute magnitudes of a group of stars to their temperature, spectral class, or color index.

**high-velocity object** An object with a high space velocity; generally, an object that does not share the high orbital velocity of the sun about the galactic nucleus.

**horizon (astronomical)** A great circle on the celestial sphere, 90° from the zenith.

**horizon system** A system of celestial coordinates (altitude and azimuth) based on the astronomical horizon and a north point.

**horizontal branch** A sequence of stars of approximately constant absolute magnitude (near $M_v = 0$) on the Hertzsprung-Russell diagram of a typical globular cluster.

**horizontal parallax** The angle by which an object on the horizon appears displaced (after a correction is made for atmospheric refraction) when it is viewed from a point on the earth's equator rather than from the center of the earth.

**hour angle** An angle measured westward along the celestial equator from a local meridian to an hour circle passing through an object.

**hour circle** A great circle on the celestial sphere passing through the celestial poles.

**Hubble constant** The constant of proportionality in the relationship between the velocities and distances of remote galaxies.

**hyperbola** A curve of intersection between a circular cone and a plane that is at too small an angle with the axis of the cone to cut all the way through it and is not parallel to a line in the face of the cone.

**image** The optical representation of an object produced by the refraction or reflection of light rays from the object by a lens or mirror.

**image tube** A device in which electrons emitted from a photocathode surface exposed to light are focused electronically.

**inclination (orbital)** The angle between the orbital plane of a revolving body and a fundamental plane—usually the plane of the celestial equator or the ecliptic.

**Index Catalogue (IC)** A supplement to Dryer's *New General Catalogue* of star clusters and nebulas.

**index of refraction** A measure of the refracting power of a transparent substance; specifically, the ratio of the speed of light in a vacuum to its speed in the substance.

**inertia** The property of matter that makes the action of a force necessary to change the state of motion of an object.

**inferior conjunction** The configuration of an inferior planet when it has the same longitude as the sun and is between the sun and the earth.

**inferior planet** A planet whose distance from the sun is less than that of the earth.

**infrared radiation** Electromagnetic radiation with a wavelength longer than the longest visible wavelength (red) but shorter than wavelengths in the radio range.

**intercalate** To insert, as a day in a calendar.

**interferometer (stellar)** An optical device that uses light-interference phenomena to measure small angles.

416

**International Date Line** An arbitrary line on the surface of the earth, near a longitude of 180°, on either side of which the date changes by one day.

**international magnitude system** A system of photographic and photovisual magnitudes based on the blue and yellow regions of the spectrum.

**interplanetary medium** Gas and solid particles in interplanetary space.

**interstellar dust** Microscopic solid grains in interstellar space.

**interstellar gas** Sparse gas in interstellar space.

**interstellar lines** Absorption lines superimposed on stellar spectra, produced by interstellar gas.

**interstellar matter** Interstellar gas and dust.

**ion** An atom that has become electrically charged through the addition or loss of one or more electrons.

**ionization** The process by which an atom gains or loses electrons.

**ionosphere** The upper region of the earth's atmosphere, in which many of the atoms are ionized.

**irregular galaxy** A galaxy without rotational symmetry (i.e., neither spiral nor elliptical).

**irregular variable** A variable star whose period of light variation is not regular.

**Jovian planet** Jupiter, Saturn, Uranus, or Neptune.

**Julian calendar** A calendar introduced by Julius Caesar in 45 B.C.

**Jupiter** The fifth planet from the sun in the solar system.

**Kepler's laws** Three laws developed by Johannes Kepler to describe the motions of the planets.

**Kirkwood's gaps** Gaps in the spacing of the periods of the minor planets due to perturbations produced by the major planets.

**latitude** A north-south coordinate on the surface of the earth; the angular distance north or south of the equator measured along a meridian.

**leap year** A year with 366 days, intercalated approximately every 4 years to make the average length of the calendar year as nearly equal as possible to the tropical year.

**libration** A change in the visible hemisphere of the moon as viewed from the earth.

**libration in latitude** A libration caused by the fact that the moon's axis of rotation is not perpendicular to its plane of revolution.

417

**libration in longitude** A libration caused by irregularity in the moon's orbital speed.

**light** Electromagnetic radiation visible to the human eye.

**light curve** A graph showing the variation in light, or magnitude, of a variable or an eclipsing binary star.

**light year** The distance light travels in a vacuum in 1 year; 6 trillion miles.

**limb** The apparent edge of a celestial body.

**limiting magnitude** The faintest magnitude that can be observed with a given instrument or under given conditions.

**line of apsides** The line connecting the apses of an orbit.

**line of nodes** The line connecting the nodes of an orbit.

**Local Group** The cluster of galaxies to which our galaxy belongs.

**longitude** An east-west coordinate on the earth's surface; the angular distance along the equator east or west of the Greenwich meridian to another meridian.

**low-velocity object** An object that has a low space velocity; generally, an object that shares the sun's high orbital speed about the galactic center.

**luminosity** The rate of radiation of energy into space by a celestial object.

**luminosity class** A subclassification of a star of a given spectral class according to its luminosity.

**lunar eclipse** An eclipse of the moon.

**Magellanic clouds** Two neighboring galaxies visible to the naked eye in southern latitudes.

**magnetic field** The region of space near a magnetized body within which magnetic forces can be detected.

**magnifying power** A measure of the strength of a telescope based on the increase in angular diameter of an object viewed through it.

**magnitude** A measure of the amount of light received from a star or another luminous object.

**main sequence** The largest sequence of stars on the Hertzsprung-Russell diagram.

**major axis (of an ellipse)** The maximum diameter of an ellipse.

**major planet** A Jovian planet.

**mare** Latin for *sea*; the name applies to many sealike features on the moon and Mars.

**Mars** The fourth planet from the sun in the solar system.

**mass** A measure of the total amount of material in a body.

**mass-luminosity relationship** An empirical relationship between the masses and luminosities of many stars.

**mass-radius relationship (of white dwarfs)** A theoretical relationship between the masses and radii of white dwarf stars.

**maximum elongation** The point in the orbit of an inferior planet at which the difference between its celestial longitude and that of the sun is greatest.

**mean solar day** The interval between successive passages of the mean sun across the meridian; the average length of the apparent solar day.

**mean solar time** Time reckoned from the position of the mean sun.

**mean sun** A fictitious body that moves eastward with uniform angular velocity along the celestial equator.

**mechanics** The branch of physics that deals with the behavior of material bodies.

**Mercury** The nearest planet to the sun in the solar system.

**meridian (celestial)** A great circle on the celestial sphere that passes through an observer's zenith and the north (or south) celestial pole.

**meridian (terrestrial)** A great circle on the surface of the earth that passes through a particular place and the North and South Poles of the earth.

**Messier Catalogue** A catalog of nonstellar objects compiled by Charles Messier in 1787.

**meteor** The luminous phenomenon observed when a meteoroid enters the earth's atmosphere and burns up; sometimes called a shooting star.

**meteor shower** The apparent descent of a large number of meteors radiating from a common point in the sky, caused by the collision of the earth with a swarm of meteoritic particles.

**meteorite** A portion of a meteoroid that survives passage through the atmosphere and strikes the ground.

**meteoroid** A meteoritic particle in space.

**Milky Way** A band of light encircling the sky, caused by the many stars lying near the plane of the galaxy; also used as a synonym for the galaxy to which the sun belongs.

**minor axis (of an ellipse)** The smallest, or least, diameter of an ellipse.

**minor planet** One of several tens of thousands of small planets, ranging from a few hundred miles to less than 1 mile in diameter.

**Mira-type variable** Any of a large

419

class of red giant long-period or irregularly pulsating variables, of which the star Mira is a prototype.

**molecule** Two or more atoms bound together.

**monochromatic** Of one wavelength, or color.

**nadir** A point on the celestial sphere 180° from the zenith.

**neap tides** The lowest tides in the month, which occur when the moon is near the first- or third-quarter phase.

**nebula** A cloud of interstellar gas or dust.

**Neptune** The eighth planet from the sun in the solar system.

**neutron** A subatomic particle with no charge and with mass approximately equal to that of the proton.

**neutron star** A star of extremely high density composed entirely of neutrons.

**New General Catalogue (NGC)** A catalog of star clusters, nebulas, and galaxies compiled by J. L. E. Dreyer in 1888.

**new moon** A phase of the moon during which its longitude is the same as that of the sun.

**Newton's laws** Laws of mechanics and gravitation formulated by Isaac Newton.

**Newtonian focus** An optical arrangement in a reflecting telescope whereby light is reflected by a flat mirror to a focus at the side of the telescope tube just before it reaches the focus of the objective lens.

**node** An intersection of the orbit of a celestial body with a fundamental plane, usually the plane of the celestial equator or the ecliptic.

**nodical month** The period of revolution of the moon about the earth with respect to the line of nodes of the moon's orbit.

**nodical (eclipse) year** The period of revolution of the earth about the sun with respect to the line of nodes of the moon's orbit.

**north point** That intersection of the celestial meridian and the astronomical horizon lying nearest the north celestial sphere.

**nova** A star that experiences a sudden outburst of radiant energy, temporarily increasing its luminosity hundreds or thousands of times.

**nuclear transformation** The transformation of one atomic nucleus into another.

**nucleus (atomic)** The heavy part of an atom, composed mostly of protons and neutrons, about which the electrons revolve.

**nucleus (cometary)** A swarm of

solid particles in the head of a comet.

**nucleus (galactic)** A concentration of stars, and possibly gas, at the center of a galaxy.

**nutation** The nodding motion of the earth's axis.

**objective lens** The principal image-forming component of a telescope or another optical instrument.

**objective prism** A prismatic lens placed in front of the objective lens of a telescope to transform each star image into an image of the stellar spectrum.

**oblate spheroid** A solid formed by rotating an ellipse about its minor axis.

**oblateness** A measure of the flattening of an oblate spheroid.

**obliquity (of the ecliptic)** The angle between the planes of the celestial equator and the ecliptic; about $23\frac{1}{2}°$.

**obscuration (interstellar)** The absorption of starlight by interstellar dust.

**occultation** An eclipse of a star or planet by the moon or another planet.

**open cluster** A comparatively loose, or "open," cluster of a few dozen to a few thousand stars in the spiral arms or disk of the galaxy; a galactic cluster.

**opposition** The configuration of a planet when its elongation is 180°.

**optical binary** Two stars at different distances that, when viewed in projection, are so nearly lined up that they appear close together, although they are not dynamically associated.

**optics** The branch of physics that deals with the properties of light.

**orbit** The path of a body revolving about another body or point.

**parabola** The curve of intersection between a circular cone and a plane parallel to a straight line in the surface of the cone.

**paraboloid** A parabola of revolution; a curved surface whose cross section is parabolic. The surface of the primary mirror in a standard reflecting telescope is a paraboloid.

**parallax** An apparent displacement of an object due to a motion of the observer.

**parallax (stellar)** An apparent displacement of a nearby star that results from the motion of the earth around the sun; numerically, the angle subtended by 1 AU at the distance of a particular star.

**parsec** The distance of an object

with a stellar parallax of 1 second of arc. A parsec equals 3.26 light years.

**partial eclipse** An eclipse in which the concealed body is not completely obscured.

**peculiar velocity** The velocity of a star with respect to the local standard of rest; that is, its space motion, corrected for the motion of the sun with respect to nearby stars.

**penumbra** The portion of a shadow from which only part of the light source is occulted by an opaque body.

**penumbral eclipse** A lunar eclipse in which the moon passes through the penumbra, but not the umbra, of the earth's shadow.

**periastron** The place in the orbit of a star in a binary star system where it is closest to its companion star.

**perigee** The place in the orbit of an earth satellite where it is closest to the center of the earth.

**perihelion** The place in the orbit of an object revolving about the sun where it is closest to the center of the sun.

**period** Generally the interval of time required for a celestial body to rotate once on its axis, revolve once about a primary, or return to its original state after an increase in luminosity.

**period-luminosity relationship** An empirical relationship between the periods and luminosities of Cepheid variable stars.

**periodic comet** A comet whose orbit has an eccentricity of less than 1.0.

**perturbation** A small disturbance in the motion of a celestial body produced by the proximity of another body.

**phases (lunar)** The changes in the moon's appearance as different portions of its illuminated hemisphere become visible from the earth.

**photocell (photoelectric cell)** An electron tube in which electrons are dislodged from the cathode when it is exposed to light and accelerated, thus producing a current in the tube whose strength serves as a measure of the light striking the cathode.

**photographic magnitude** The magnitude of an object, as measured on traditional blue- and violet-sensitive photographic emulsions.

**photometry** The measurement of light intensities.

**photomultiplier** A photoelectric cell in which the electric current generated is amplified at several stages within the tube.

**photon** A discrete unit of electromagnetic energy.

**photosphere** The region of the solar (or stellar) atmosphere from which radiation escapes into space.

**photovisual magnitude** A magnitude corresponding to the spectral region to which the human eye is most sensitive but measured photographically with green- and yellow-sensitive emulsions and filters.

**plage** A bright region of the solar surface observed in the monochromatic light of a particular spectral line; a flocculus.

**Planck's law of radiation** A formula for calculating the intensity of radiation at various wavelengths emitted by a blackbody.

**planet** Any of nine solid, nonluminous bodies revolving about the sun.

**planetary nebula** A shell of gas ejected from, and enlarging about, an extremely hot star.

**planetoid** A minor planet.

**Pluto** The ninth planet from the sun in the solar system.

**polar axis** The axis of rotation of a planet; also, an axis in the mounting of a telescope that is parallel to the earth's axis.

**position angle** The direction in the sky of one celestial object with respect to another; for example, the angle, measured to the east from the north, showing the position of the fainter component of a visual binary star with respect to the brighter component.

**precession (of the equinoxes)** The slow westward motion of the equinoxes along the ecliptic due to precession.

**precession (terrestrial)** A slow, conical motion of the earth's axis of rotation, caused principally by the gravitational torque of the moon and sun on the earth's equatorial bulge.

**primary minimum (in the light curve of an eclipsing binary)** The middle of the eclipse, during which the most light is lost.

**prime focus** The point in a telescope where the objective lens focuses the light.

**prime meridian** The meridian of longitude passing through the site of the old Royal Greenwich Observatory, near London; the great circle from which terrestrial longitude is measured.

**primeval atom** A single mass whose explosion (in some cosmologies) began the expansion of the universe.

**primeval fireball** The extremely hot opaque gas presumed to have con-

423

tained the entire mass of the universe at the time of or immediately following the explosion of the primeval atom.

**prism** A wedge-shaped piece of glass used to disperse white light into a spectrum.

**prominence** A flamelike phenomenon in the solar corona.

**proper motion** The angular change in direction of a star during one year.

**proton** A heavy subatomic particle that carries a positive charge; one of the two principal constituents of the atomic nucleus.

**pulsar** A small but powerful celestial radio source with very regular, short periods.

**pulsating variable** A variable star that periodically changes in size and luminosity.

**quadrature** A configuration of a planet in which its elongation is 90°.

**quarter moon** Either of the two phases of the moon when its longitude differs by 90° from that of the sun and it appears half full.

**quasar** A quasi-stellar source.

**quasi-stellar galaxy (QSG)** An apparently stellar object with a very large red shift, presumed to be extragalactic and highly luminous.

**quasi-stellar source (QSS)** An apparently stellar object with a very large red shift that is a strong source of radio waves, presumed to be extragalactic and highly luminous.

**radial velocity** The component of relative velocity that lies in the line of sight.

**radial-velocity curve** A plot of the variation in radial velocity over time of a binary or variable star.

**radiant (of meteor shower)** The point in the sky from which the meteors belonging to a shower seem to radiate.

**radiation** The emission and transmission of energy in the form of waves or particles.

**radio astronomy** The use of radio wavelengths to make astronomical observations.

**radio telescope** A telescope designed to make observations in radio wavelengths.

**ray (lunar)** Any of a system of bright, elongated streaks, sometimes associated with a crater.

**recurrent nova** A nova that has erupted more than once.

**red giant** A large, cool star of high luminosity in the upper right portion of the Hertzsprung-Russell diagram.

**red shift** A shift to longer wavelengths of the light from remote galaxies, presumably caused by their velocity of recession.

**reddening (interstellar)** A change in the color, or wavelength, of starlight passing through interstellar dust, which scatters blue light more effectively than red.

**reflecting telescope** A telescope in which the principal optical component is a concave mirror.

**reflection nebula** A relatively dense interstellar dust cloud that is illuminated by starlight.

**refracting telescope** A telescope in which the principal optical component (objective) is a lens or system of lenses.

**refraction** The bending of light rays passing from one transparent medium to another.

**regression of nodes** A consequence of certain perturbations of the orbit of a revolving body whereby the nodes of the orbit slide westward in the fundamental plane (usually the plane of the ecliptic or the celestial equator.)

**relative orbit** The orbit of one of two mutually revolving bodies about the other.

**resolution** The degree to which fine details in an image are separated, or resolved.

**resolving power** A measure of the ability of an optical system to resolve, or separate, fine details in the image it produces.

**retrograde motion** An apparent westward motion of a planet on the celestial sphere or with respect to the stars.

**revolution** The motion of one body around another.

**right ascension** A coordinate for measuring the east-west positions of celestial bodies; the angle, measured eastward along the celestial equator, from the vernal equinox to the hour circle passing through a body.

**rill (lunar)** A crevasse, or trench-like depression, in the moon's surface.

**rotation** The turning of a body about an axis running through it.

**RR Lyrae variable** One of a class of giant pulsating stars with periods of less than 1 day; a cluster variable.

**saros** A particular cycle of similar eclipses that recur at intervals of about 18 years.

**satellite** A body, such as the earth's moon, that revolves about a larger body.

**Saturn** The sixth planet from the sun in the solar system.

**Schmidt telescope** A reflecting telescope in which certain aberrations produced by a spherical concave mirror are compensated for by a thin objective correcting lens.

**secondary minimum (in an eclipsing binary's light curve)** The middle of the eclipse of the fainter star by the brighter, at which time the light of the system diminishes less than it does during the eclipse of the brighter star by the fainter.

**secular** Not periodic.

**secular parallax** A mean parallax for a selection of stars, derived from the components of their proper motions that reflect the motion of the sun.

**seleno-** A prefix referring to the moon.

**semimajor axis** Half the major axis of a conic section.

**separation (in a visual binary)** The angular separation of the two components of a visual binary star.

**shadow cone** The umbra of the shadow of a spherical body (such as the earth) in sunlight.

**shell star** A type of star, usually a Class B, A, or F star, surrounded by a gaseous ring, or shell.

**sidereal day** The interval between two successive meridian passages of the vernal equinox.

**sidereal month** The period of the moon's revolution about the earth with respect to the stars.

**sidereal period** The period of revolution of one body about another with respect to the stars.

**sidereal time** The local hour angle of the vernal equinox.

**sidereal year** The period of the earth's revolution about the sun with respect to the stars.

**small circle** Any circle on the surface of a sphere that is not a great circle.

**solar activity** Phenomena in the solar atmosphere associated with sunspots, plages, and the like.

**solar antapex** The direction away from which the sun is moving with respect to the local standard of rest.

**solar apex** The direction toward which the sun is moving with respect to the local standard of rest.

**solar motion** The motion, or velocity, of the sun with respect to the local standard of rest.

**solar parallax** The angle subtended by the equatorial radius of the earth at a distance of 1 AU.

**solar system** The system of the sun and the planets, satellites, asteroids, comets, meteoroids, and other objects revolving around it.

**solar time** Time measured according to the sun.

426

**solar wind**  A radial flow of corpuscular radiation leaving the sun.

**solstice**  Either of two points on the celestial sphere where the sun reaches its maximum distances north or south of the celestial equator.

**south point**  The intersection of the celestial meridian and the astronomical horizon 180° from the north point.

**space motion**  The velocity of a star with respect to the sun.

**spectral class (or type)**  A classification of a star according to the characteristics of its spectrum.

**spectral sequence**  The sequence of spectral classes of stars arranged in order of decreasing temperature.

**spectrogram**  A photograph of a spectrum.

**spectrograph**  An instrument for photographing a spectrum; usually attached to a telescope to photograph the spectrum of a star.

**spectroheliogram**  A photograph of the sun obtained with a spectroheliograph.

**spectroheliograph**  An instrument for photographing the sun, or part of the sun, in the monochromatic light of a particular spectral line.

**spectroscope**  An instrument for viewing the spectrum of a light source directly.

**spectroscopic binary star**  A binary star in which the components are not resolvable optically, but whose binary nature is indicated by periodic variations in its radial velocity indicating orbital motion.

**spectroscopic parallax**  A parallax derived by comparing the apparent magnitude of the star with its absolute magnitude as deduced from its spectral characteristics.

**spectroscopy**  The study of spectra.

**spectrum**  The array of colors, or wavelengths, obtained when light from a source is dispersed by passing it through a prism or grating.

**spectrum analysis**  The study and analysis of spectra, especially stellar spectra.

**spectrum binary**  A star whose binary nature is revealed by spectral characteristics that can only result from a composite of the spectra of two different stars.

**spicule**  A narrow jet of material rising in the solar chromosphere.

**spiral arms**  Armlike areas of interstellar material and young stars that wind out in a plane from the central nucleus of a spiral galaxy.

**spiral galaxy**  A flattened, rotating galaxy with wheellike arms of interstellar material and young stars winding out from its nucleus.

**spring tide** The highest tide of the month, produced when the longitudes of the sun and moon differ from each other by nearly 0° or 180°.

**standard time** The local mean solar time of a standard meridian, adopted over a large region to avoid the inconvenience of continuous time changes around the earth.

**star** A self-luminous sphere of gas.

**star cluster** An assemblage of stars held together by their mutual gravitation.

**statistical parallax** The mean parallax of a selection of stars, derived from the radial velocities of the stars and the components of their proper motions that cannot be affected by the solar motion.

**steady-state theory** A cosmological theory based on the cosmological principle and the continuous creation of matter.

**Stefan-Boltzmann law** A formula for computing the rate at which a blackbody radiates energy.

**stellar evolution** The changes that take place in the size, luminosity, structure, and other characteristics of a star as it ages.

**stellar parallax** The angle subtended by 1 AU at the distance of a star, usually measured in seconds of arc.

**subdwarf** A star less luminous than main-sequence stars of the same spectral type.

**subgiant** A star more luminous than main-sequence stars of the same spectral type but less so than normal giants of that type.

**summer solstice** The point on the celestial sphere where the sun is farthest north of the celestial equator.

**sun** The star about which the earth and other planets revolve.

**sunspot** A temporary cool region in the solar photosphere that appears dark in contrast to the surrounding hotter photosphere.

**sunspot cycle** A semiregular 11-year period during which the frequency of sunspots fluctuates.

**supergiant** A star of very high luminosity.

**superior conjunction** The configuration of a planet in which it has the same longitude as the sun but is more distant than the sun.

**superior planet** A planet farther from the sun than the earth.

**supernova** A stellar outburst, or explosion, in which a star suddenly increases in luminosity by hundreds of thousands or hundreds of millions of times.

**surface gravity** The acceleration of

gravity at the surface of a planetary body.

**synodic month** The period of revolution of the moon with respect to the sun; or the period of the cycle of lunar phases.

**synodic period** The interval between successive occurrences of the same configuration of a planet (for example, between successive oppositions or successive superior conjunctions).

**tail (cometary)** Gases and solid particles ejected from the head of a comet and forced away from the sun by radiation pressure or corpuscular radiation.

**tangential (transverse) velocity** The component of a star's space velocity that lies in the plane of the sky.

**tektites** Rounded glassy bodies suspected to be of meteoritic origin.

**telescope** An optical instrument used to view, or measure, or photograph distant objects.

**telluric** Of terrestrial origin.

**temperature (absolute)** Temperature measured in centigrade degrees from absolute zero.

**temperature (centigrade)** Temperature measured on a scale calibrated so that water freezes at 0° and boils at 100°.

**temperature (color)** The temperature of a star as estimated from the intensity of the stellar radiation in two or more colors, or wavelengths.

**temperature (effective)** The temperature of a blackbody that would radiate the same total amount of energy as a particular body.

**temperature (Fahrenheit)** Temperature measured on a scale calibrated so that water freezes at 32° and boils at 212°.

**temperature (Kelvin)** Absolute temperature measured in centigrade degrees (Kelvin temperature = centigrade temperature + 273°).

**temperature (radiation)** The temperature of a blackbody that radiates the same amount of energy in a given spectral region as a particular body.

**terminator** The line of sunrise or sunset on a celestial body such as the moon.

**terrestrial planet** Mercury, Venus, Earth, Mars, and sometimes Pluto.

**thermonuclear energy** Energy associated with thermonuclear reactions.

**thermonuclear reaction** A nuclear reaction or transformation that results from encounters between high-velocity nuclear particles.

**tidal force** A differential gravitational force that tends to deform a body.

**tide**  A deformation of a body caused by the differential gravitational force exerted on it by another body.

**total eclipse**  An eclipse of the sun in which the photosphere is entirely hidden by the moon; a lunar eclipse in which the moon passes completely into the umbra of the earth's shadow.

**train (of a meteor)**  A temporarily luminous trail in the wake of a meteor.

**triangulation**  The measurement of some of the elements of a triangle so that other elements can be calculated by trigonometric operations; a method of determining distances without taking direct measurements.

**Trojan minor planet**  One of several minor planets that share Jupiter's orbit around the sun but are located approximately 60° around the orbit from Jupiter.

**tropical year**  The period of revolution of the earth about the sun with respect to the vernal equinox.

**UBV system**  A system of measuring stellar magnitudes in the ultraviolet, blue, and green-yellow regions of the spectrum.

**ultraviolet radiation**  Electromagnetic radiation whose wavelength is shorter than the shortest wavelengths to which the eye is sensitive (violet); radiation whose wavelength ranges from approximately 100 Å to 4000 Å.

**umbra**  The central, completely dark part of a shadow.

**universal time**  The local mean time of the prime meridian.

**universe**  The totality of matter, radiation, and space.

**Uranus**  The seventh planet from the sun in the solar system.

**Venus**  The second planet from the sun in the solar system.

**vernal equinox**  The point on the celestial sphere where the sun crosses from the south to the north of the celestial equator.

**vertical circle**  Any great circle passing through the zenith.

**visual binary star**  A binary star in which the two components can be resolved telescopically.

**visual photometer**  An instrument used with a telescope to measure, visually, the light flux from a star.

**volume**  A measure of the total space occupied by a body.

**walled plain**  A large lunar crater.

**wandering (of the poles)**  A semiperiodic shift of the body of the earth relative to its axis of rotation.

**wavelength** The spacing of the crests or troughs in a wave train.

**weight** A measure of the force of gravitational attraction.

**west point** The point on the horizon 270° from the north point, measured clockwise from the zenith.

**white dwarf** A star that has exhausted most or all of its nuclear fuel and has collapsed to a very small size. White dwarfs are believed to be stars near the final stage of evolution.

**Widmanstätten figures** Crystalline structures observable in cut and polished meteorites.

**Wien's law** A formula relating the temperature of a blackbody to the exact wavelength at which it emits the greatest intensity of radiation.

**winter solstice** The point on the celestial sphere where the sun is furthest south of the celestial equator.

**Wolf-Rayet star** One of a class of very hot stars that eject shells of gas at a very high velocity.

**x-ray stars** Stars that emit observable amounts of radiation at x-ray frequencies.

**x-rays** Photons whose wavelengths are shorter than ultraviolet wavelengths and longer than gamma wavelengths.

**Zeeman effect** A splitting or broadening of spectral lines due to magnetic fields.

**zenith** The point on the celestial sphere opposite the direction of gravity; the direction opposite to that indicated by a plumb bob.

**zenith distance** The distance, in degrees of arc, of a point on the celestial sphere from the zenith.

**zodiac** A belt around the sky centered on the ecliptic.

**zodiacal light** A faint illumination along the zodiac, believed to be due to sunlight reflected and scattered by interplanetary dust.

**zone of avoidance** A region near the Milky Way where the interstellar dust is so thick that few or no exterior galaxies can be seen.

**zone time** The time, kept in a zone 15° wide, equal to the local mean time of the central meridian of the zone.

# APPENDIX 1
# SUMMARY TABLES

# 1. THE PLANETS (PHYSICAL DATA)

| Planet | Diameter (miles) | Diameter (Earth = 1) | Mass (Earth = 1) | Surface Gravity (Earth = 1) | Period of Rotation | Number of Moons |
|---|---|---|---|---|---|---|
| Mercury | 3,025 | 0.38 | 0.06 | 0.38 | 58.65 days | 0 |
| Venus | 7,526 | 0.95 | 0.82 | 0.90 | 243 days | 0 |
| Earth | 7,927 | 1.00 | 1.00 | 1.00 | 23 hr 56 min | 1 |
| Mars | 4,218 | 0.53 | 0.11 | 0.38 | 24 hr 37 min | 2 |
| Jupiter | 88,700 | 11.19 | 318.0 | 2.64 | 9 hr 50 min | 12 |
| Saturn | 75,100 | 9.47 | 95.2 | 1.13 | 10 hr 14 min | 10 |
| Uranus | 29,200 | 3.69 | 14.6 | 1.07 | 10 hr 49 min | 5 |
| Neptune | 31,650 | 3.50 | 17.3 | 1.08 | 16 hr | 2 |
| Pluto | 3,500? | <.5 | 0.1? | 0.3? | 6.39 days | 0 |

# 2. THE PLANETS (ORBITAL DATA)

| Planet | Average Distance from the Sun (AU's) | Average Distance from the Sun (millions of miles) | Orbital Period (years) | Orbital Period (days) | Orbital Speed (mi./sec.) | Orbital Inclination |
|---|---|---|---|---|---|---|
| Mercury | 0.387 | 36.0 | 0.241 | 88.0 | 29.7 | 7.0 |
| Venus | 0.723 | 67.2 | 0.615 | 224.7 | 21.8 | 3.4 |
| Earth | 1.000 | 92.9 | 1.000 | 365.3 | 18.5 | 0.0 |
| Mars | 1.524 | 141.5 | 1.881 | 687.0 | 15.0 | 1.8 |
| Jupiter | 5.203 | 483.4 | 11.862 | | 8.1 | 1.3 |
| Saturn | 9.539 | 886.0 | 29.458 | | 6.0 | 2.5 |
| Uranus | 19.18 | 1782.0 | 84.013 | | 4.2 | 0.8 |
| Neptune | 30.06 | 2792.0 | 164.793 | | 3.4 | 1.8 |
| Pluto | 39.44 | 3664.0 | 247.686 | | 2.9 | 17.2 |

# 3. SATELLITES OF PLANETS

| Name | Maximum Magnitude | Diameter (miles) | Average Distance from Planet (miles) | Period of Revolution | Discoverer |
|------|-------------------|------------------|--------------------------------------|----------------------|------------|
| **Satellite of Earth** | | | | | |
| Moon | −12.7 | 2160 | 238,900 | 27$^d$ 07$^h$ 43$^m$ | |
| **Satellites of Mars** | | | | | |
| Phobos | 11.6 | 12 | 5,800 | 0$^d$ 07$^h$ 39$^m$ | Hall, 1877 |
| Deimos | 12.8 | (<10)* | 14,600 | 1  06  18 | Hall, 1877 |
| **Satellites of Jupiter** | | | | | |
| V | 13.0 | (100) | 112,000 | 0$^d$ 11$^h$ 57$^m$ | Barnard, 1892 |
| Io | 4.8 | 2020 | 262,000 | 1  18  28 | Galileo, 1610 |
| Europa | 5.2 | 1790 | 417,000 | 3  13  14 | Galileo, 1610 |
| Ganymede | 4.5 | 3120 | 665,000 | 7  03  43 | Galileo, 1610 |
| Callisto | 5.5 | 2770 | 1,171,000 | 16  16  32 | Galileo, 1610 |
| VI | 13.7 | (50) | 7,133,000 | 250  14 | Perrine, 1904 |
| VII | 16 | (20) | 7,295,000 | 259  16 | Perrine, 1905 |
| X | 18.6 | (<10) | 7,369,000 | 263  13 | Nicholson, 1938 |
| XII | 18.8 | (<10) | 13,200,000 | 631  02 | Nicholson, 1951 |
| XI | 18.1 | (<10) | 14,000,000 | 692  12 | Nicholson, 1938 |
| VIII | 18.8 | (<10) | 14,600,000 | 738  22 | Melotte, 1908 |
| IX | 18.3 | (<10) | 14,700,000 | 758 | Nicholson, 1914 |
| **Satellites of Saturn** | | | | | |
| Janus | (14) | <300 | 100,000 | 0$^d$ 17$^h$ 59$^m$ | Dollfus, 1966 |
| Mimas | 12.1 | 300 | 116,000 | 0  22  37 | Herschel, 1789 |
| Enceladus | 11.8 | 400 | 148,000 | 1  08  53 | Herschel, 1789 |
| Tethys | 10.3 | 600 | 183,000 | 1  21  18 | Cassini, 1684 |
| Dione | 10.4 | 600 | 235,000 | 2  17  41 | Cassini, 1684 |
| Rhea | 9.8 | 810 | 327,000 | 4  12  25 | Cassini, 1672 |
| Titan | 8.4 | 2980 | 759,000 | 15  22  41 | Huygens, 1655 |
| Hyperion | 14.2 | (100) | 920,000 | 21  06  38 | Bond, 1848 |
| Iapetus | 11.0 | (500) | 2,213,000 | 79  07  56 | Cassini, 1671 |
| Phoebe | (14) | (100) | 8,053,000 | 550  11 | Pickering, 1898 |
| **Satellites of Uranus** | | | | | |
| Miranda | 16.5 | (200) | 77,000 | 1$^d$ 09$^h$ 56$^m$ | Kuiper, 1948 |
| Ariel | 14.4 | (500) | 119,000 | 2  12  29 | Lassell, 1851 |
| Umbriel | 15.3 | (300) | 166,000 | 4  03  38 | Lassell, 1851 |
| Titania | 14.0 | (600) | 272,000 | 8  16  56 | Herschel, 1787 |
| Oberon | 14.2 | (500) | 365,000 | 13  11  07 | Herschel, 1787 |
| **Satellites of Neptune** | | | | | |
| Triton | 13.6 | 2300 | 220,000 | 5$^d$ 21$^h$ 03$^m$ | Lassell, 1846 |
| Nereid | 18.7 | (200) | 3,461,000 | 359  10 | Kuiper, 1949 |

*Parentheses indicate that diameters are estimated.

# 4. THE BRIGHTEST STARS

| | Name | Common Name | Apparent Magnitude | Position (1970) R.A. | Decl. | Distance (light yrs.) |
|---|---|---|---|---|---|---|
| 1. | $\alpha$ Canis Majoris | Sirius | −1.58 | 6ʰ 44ᵐ | −16° 41′ | 8.7 |
| 2. | $\alpha$ Carinae | Canopus | −0.86 | 6  23 | −52  41 | 98 |
| 3. | $\alpha$ Centauri | Rigil Kentaurus | 0.06 | 14  38 | −60  43 | 4.3 |
| 4. | $\alpha$ Lyrae | Vega | 0.14 | 18  36 | +38  45 | 26.5 |
| 5. | $\alpha$ Aurigae | Capella | 0.21 | 5  14 | +45  58 | 45 |
| 6. | $\alpha$ Bootis | Arcturus | 0.24 | 14  14 | +19  20 | 36 |
| 7. | $\beta$ Orionis | Rigel | 0.34 | 5  13 | −08  14 | 900 |
| 8. | $\alpha$ Canis Minoris | Procyon | 0.48 | 7  38 | +05  18 | 11.3 |
| 9. | $\alpha$ Eridani | Achernar | 0.60 | 1  37 | −57  23 | 118 |
| 10. | $\beta$ Centauri | Agena | 0.86 | 14  02 | −60  13 | 490 |
| 11. | $\alpha$ Aquilae | Altair | 0.89 | 19  49 | +08  47 | 16.5 |
| 12. | $\alpha$ Orionis | Betelgeuse | 0.92 (var.) | 5  54 | +07  24 | 520 |
| 13. | $\alpha$ Crucis | Acrux | 1.05 | 12  25 | −62  56 | 370 |
| 14. | $\alpha$ Tauri | Aldebaran | 1.06 | 4  34 | +16  27 | 68 |
| 15. | $\beta$ Geminorum | Pollux | 1.21 | 7  44 | +28  06 | 35 |
| 16. | $\alpha$ Virginis | Spica | 1.21 | 13  24 | −11  00 | 220 |
| 17. | $\alpha$ Scorpii | Antares | 1.22 | 16  28 | −26  22 | 520 |
| 18. | $\alpha$ Piscis Austrini | Fomalhaut | 1.29 | 22  56 | −29  47 | 22.6 |
| 19. | $\alpha$ Cygni | Deneb | 1.33 | 20  40 | +45  10 | 1600 |
| 20. | $\alpha$ Leonis | Regulus | 1.34 | 10  07 | +12  07 | 84 |
| 21. | $\beta$ Crucis | | 1.50 | 12  46 | −59  32 | 490 |
| 22. | $\alpha$ Geminorum | Castor | 1.58 | 7  33 | +31  57 | 45 |
| 23. | $\gamma$ Crucis | Gacrux | 1.61 | 12  30 | −56  57 | 220 |
| 24. | $\epsilon$ Canis Majoris | Adhara | 1.63 | 6  57 | −28  56 | 680 |
| 25. | $\epsilon$ Ursae Majoris | Alioth | 1.68 | 12  53 | +56  07 | 68 |
| 26. | $\gamma$ Orionis | Bellatrix | 1.70 | 5  24 | +06  19 | 470 |
| 27. | $\lambda$ Scorpii | Shaula | 1.71 | 17  32 | −37  05 | 310 |
| 28. | $\epsilon$ Carinae | Avior | 1.74 | 8  22 | −59  24 | 340 |
| 29. | $\epsilon$ Orionis | Alnilam | 1.75 | 5  35 | −01  13 | 1600 |
| 30. | $\beta$ Tauri | El Nath | 1.78 | 5  24 | +28  35 | 300 |
| 31. | $\beta$ Carinae | Miaplacidus | 1.80 | 9  13 | −69  36 | 86 |
| 32. | $\alpha$ Trianguli Australis | Atria | 1.88 | 16  46 | −68  59 | 82 |
| 33. | $\alpha$ Persei | Marfak | 1.90 | 3  22 | +49  45 | 570 |
| 34. | $\eta$ Ursae Majoris | Alkaid | 1.91 | 13  46 | +49  28 | 210 |
| 35. | $\gamma$ Velorum | Regor | 1.92 | 8  09 | −47  16 | 520 |
| 36. | $\gamma$ Geminorum | Alhena | 1.93 | 6  36 | +16  26 | 105 |
| 37. | $\alpha$ Ursae Majoris | Dubhe | 1.95 | 11  02 | +61  55 | 105 |
| 38. | $\epsilon$ Sagittarii | Kaus Australis | 1.95 | 18  22 | −34  24 | 124 |
| 39. | $\delta$ Canis Majoris | Wezen | 1.98 | 7  07 | −26  21 | 2100 |
| 40. | $\beta$ Canis Majoris | Mirzam | 1.99 | 6  21 | −17  56 | 750 |

## 4. THE BRIGHTEST STARS (Continued)

| Name | Common Name | Apparent Magnitude | Position (1970) R.A. | Decl. | Distance (light yrs.) |
|------|-------------|--------------------|----------------------|-------|-----------------------|
| 41. δ Velorum | | 2.01 | 8  44 | −54  36 | 76 |
| 42. θ Scorpii | Sargas | 2.04 | 17  35 | −42  59 | 650 |
| 43. ζ Orionis | Alnitak | 2.05 | 5  39 | −01  57 | 1600 |
| 44. β Aurigae | Menkalinan | 2.07 | 5  57 | +44  57 | 88 |
| 45. α Cassiopeiae | Schedar | 2.1  (var.) | 00  39 | +56  22 | 150 |
| 46. α Pavonis | Peacock | 2.12 | 20  23 | −56  50 | 310 |
| 47. α Ursae Minoris | Polaris | 2.12 | 2  02 | +89  08 | 680 |
| 48. σ Sagittarii | Nunki | 2.14 | 18  53 | −26  20 | 300 |
| 49. α Ophiuchi | Rasalhague | 2.14 | 17  34 | +12  35 | 58 |
| 50. α Andromedae | Alpheratz | 2.15 | 00  07 | +28  55 | 90 |

# 5. THE NEAREST STARS

| Name | Position (1970) R.A. | Position (1970) Decl. | Distance (light years) | Apparent Magnitude |
|---|---|---|---|---|
| α Centauri* | 14ʰ 37ᵐ | −60° 43′ | 4.3 | 0.1 |
| Barnard's Star | 17 56 | +04 36 | 5.9 | 9.5 |
| Wolf 359 | 10 55 | +07 13 | 7.6 | 13.5 |
| Lalande 21185 | 11 02 | +36 10 | 8.1 | 7.5 |
| Sirius* | 6 44 | −16 41 | 8.6 | −1.5 |
| Luyten 726−8* | 1 37 | −18 07 | 8.9 | 12.5 |
| Ross 154 | 18 48 | −23 51 | 9.4 | 10.6 |
| Ross 248 | 23 40 | +44 01 | 10.3 | 12.2 |
| ε Eridani | 3 32 | −09 34 | 10.7 | 3.7 |
| Luyten 789−6 | 22 37 | −15 31 | 10.8 | 12.2 |
| Ross 128 | 11 46 | +01 01 | 10.8 | 11.1 |
| 61 Cygni* | 21 06 | +38 36 | 11.2 | 5.2 |
| ε Indi | 22 02 | −56 55 | 11.2 | 4.7 |
| Procyon* | 7 38 | +05 18 | 11.4 | 0.3 |
| Σ 2398* | 18 42 | +59 35 | 11.5 | 8.9 |
| Groom. 34* | 00 17 | +43 51 | 11.6 | 8.1 |
| Lacaille 9352 | 23 04 | −36 02 | 11.7 | 7.4 |
| τ Ceti | 1 43 | −16 06 | 11.9 | 3.5 |
| BD +5° 1668 | 7 26 | +05 28 | 12.2 | 9.8 |
| Lacaille 8760 | 21 15 | −39 00 | 12.5 | 6.7 |
| Kapteyn's Star | 5 11 | −45 00 | 12.7 | 8.8 |
| Kruger 60* | 22 27 | +57 33 | 12.8 | 9.7 |
| Ross 614* | 6 28 | −02 48 | 13.1 | 11.3 |
| BD −12° 4523 | 16 29 | −12 35 | 13.1 | 10.0 |
| van Maanen's Star | 00 47 | +05 16 | 13.9 | 12.4 |
| Wolf 424* | 12 32 | +09 12 | 14.2 | 12.6 |
| CD −37° 15492 | 00 03 | −37 30 | 14.5 | 8.6 |
| Groom. 1618 | 10 09 | +49 36 | 15.0 | 6.6 |

An asterisk (*) indicates that this star is actually double or multiple. In such cases the magnitude of the brightest component is given.

# 6. THE BRIGHTEST GALAXIES

| Name | Position (1970) | | Apparent Magnitude | Distance (thousands of light years) |
|---|---|---|---|---|
| | R.A. | Decl. | | |
| LMC | 5$^h$ 23$^m$8 | −69° 47′ | 0.9 | 160 |
| SMC | 00 51.7 | −72 59 | 2.9 | 190 |
| Ursa Minor | 15 08.4 | +67 13 | | 250 |
| Draco | 17 19.7 | +57 57 | | 260 |
| Sculptor | 00 58.4 | −33 52 | 10.5 | 280 |
| Fornax | 2 38.3 | −34 39 | 9.1 | 430 |
| Leo I | 10 06.9 | +12 27 | 11.3 | 750 |
| Leo II | 11 11.9 | +22 19 | 12.8 | 750 |
| NGC 6822 | 19 43.2 | −14 50 | 9.2 | 1700 |
| M31 | 00 41.1 | +41 07 | 4.3 | 2100 |
| NGC 205 | 00 38.7 | +41 32 | 8.9 | 2100 |
| M32 | 00 41.1 | +40 43 | 9.1 | 2100 |
| NGC 185 | 00 37.2 | +48 11 | 10.3 | 2100 |
| NGC 147 | 00 31.5 | +48 11 | 10.6 | 2100 |
| M33 | 1 32.2 | +30 30 | 6.2 | 2400 |
| IC 1613 | 1 03.5 | +01 58 | 10.0 | 2400 |

# 7. THE NEAREST GALAXIES

| | Name | Position (1970) | | Apparent Magnitude | Distance (millions of light years) |
|---|---|---|---|---|---|
| | | R.A. | Decl. | | |
| 1. | LMC | 5$^h$ 23$^m$.8 | −69° 47′ | 0.9 | 0.2 |
| 2. | SMC | 00 51.7 | −72 59 | 2.9 | 0.2 |
| 3. | M31 | 00 41.1 | +41 07 | 4.3 | 2.1 |
| 4. | M33 | 1 32.2 | +30 30 | 6.2 | 2.4 |
| 5. | M83 | 13 35.4 | −29 43 | 7.0 | 8.0 |
| 6. | NGC 253 | 00 46.1 | −25 27 | 7.0 | 7.5 |
| 7. | M81 | 9 53.1 | +69 12 | 7.8 | 6.5 |
| 8. | NGC 5128 | 13 23.6 | −42 51 | 7.9 | |
| 9. | NGC 55 | 00 13.5 | −39 23 | 7.9 | 7.5 |
| 10. | NGC 4945 | 13 03.5 | −49 19 | 8.0 | |
| 11. | M101 | 14 02.1 | +54 29 | 8.2 | 14.0 |
| 12. | NGC 300 | 00 53.5 | −37 51 | 8.7 | 7.5 |
| 13. | NGC 2403 | 7 33.9 | +65 40 | 8.8 | 6.5 |
| 14. | M51 | 13 28.6 | +47 21 | 8.9 | 14.0 |
| 15. | NGC 205 | 00 38.7 | +41 32 | 8.9 | 2.1 |
| 16. | NGC 4258 | 12 17.5 | +47 28 | 8.9 | 14.0 |
| 17. | M94 | 12 49.5 | +41 16 | 8.9 | 14.0 |
| 18. | M32 | 00 41.1 | +40 43 | 9.1 | 2.1 |
| 19. | Fornax | 2 38.3 | −34 39 | 9.1 | 0.4 |
| 20. | M104 | 12 38.3 | −11 28 | 9.2 | 37.0 |
| 21. | M82 | 9 53.6 | +69 50 | 9.2 | 6.5 |
| 22. | NGC 6822 | 19 43.2 | −14 50 | 9.2 | 1.7 |
| 23. | M63 | 13 14.4 | +42 11 | 9.3 | 14.0 |
| 24. | M64 | 12 55.3 | +21 51 | 9.3 | 12.0 |
| 25. | M49 | 12 28.3 | +08 09 | 9.3 | 37.0 |
| 26. | NGC 247 | 00 45.6 | −20 54 | 9.5 | 7.5 |
| 27. | NGC 2903 | 9 30.4 | +21 39 | 9.5 | 19.0 |

# 8. MESSIER CATALOGUE

| M | NGC | Position (1970) R.A. | Decl. | Const. | App. Mag. | Distance (light yrs.) | Description |
|---|---|---|---|---|---|---|---|
| 1 | 1952 | 5ʰ 32ᵐ7 | +22° 01′ | Tau | 10 | 3,500 | "Crab nebula," supernova of 1054 |
| 2 | 7089 | 21 31.9 | −00 57 | Agr | 7 | 45,000 | Globular cluster |
| 3 | 5272 | 13 40.8 | +28 32 | CVn | 6 | 40,000 | Globular cluster |
| 4 | 6121 | 16 21.8 | −26 26 | Sco | 6 | 10,000 | Globular cluster |
| 5 | 5904 | 15 17.0 | +02 13 | Ser | 6 | 30,000 | Globular cluster |
| 6 | 6405 | 17 38.1 | −32 11 | Sco | 6 | 1,800 | Open cluster |
| 7 | 6475 | 17 51.9 | −34 48 | Sco | 5 | 800 | Bright open cluster |
| 8 | 6523 | 18 01.8 | −24 23 | Sgr | | 5,000 | "Lagoon nebula" |
| 9 | 6333 | 17 17.5 | −18 29 | Oph | 7 | 26,000 | Globular cluster |
| 10 | 6254 | 16 55.5 | −04 04 | Oph | 7 | 23,000 | Globular cluster |
| 11 | 6705 | 18 49.5 | −06 19 | Sct | 6 | 5,500 | Open cluster |
| 12 | 6218 | 16 45.6 | −01 54 | Oph | 7 | 23,000 | Globular cluster |
| 13 | 6205 | 16 40.6 | +36 31 | Her | 6 | 26,000 | The "great" globular cluster |
| 14 | 6402 | 17 36.0 | −03 14 | Oph | 8 | 23,000 | Globular cluster |
| 15 | 7078 | 21 28.6 | +12 02 | Peg | 6 | 40,000 | Globular cluster |
| 16 | 6611 | 18 17.2 | −13 48 | Ser | 7 | 8,000 | Open cluster |
| 17 | 6618 | 18 19.1 | −16 12 | Sgr | | 5,000 | "Omega" or "Horseshoe" nebula |
| 18 | 6613 | 18 18.2 | −17 09 | Sgr | 7 | 5,000 | Open cluster |
| 19 | 6273 | 17 00.2 | −26 13 | Oph | 7 | 23,000 | Globular cluster |
| 20 | 6514 | 18 00.6 | −23 02 | Sgr | | 4,000 | "Trifid nebula" |
| 21 | 6531 | 18 02.8 | −22 30 | Sgr | 7 | 4,000 | Open cluster |
| 22 | 6656 | 18 34.6 | −23 56 | Sgr | 6 | 10,000 | Globular cluster |
| 23 | 6494 | 17 55.1 | −19 00 | Sgr | 7 | 2,100 | Open cluster |
| 24 | 6603 | 18 16.7 | −18 27 | Sgr | 6 | 9,000 | Star cloud |
| 25 | IC 4725 | 18 29.9 | −19 16 | Sgr | | 2,000 | Open cluster |
| 26 | 6694 | 18 43.6 | −09 26 | Sct | 8 | 5,000 | Open cluster |
| 27 | 6853 | 19 58.4 | +22 38 | Vul | 8 | 700 | "Dumbbell nebula" (planetary) |
| 28 | 6626 | 18 22.6 | −24 52 | Sgr | 8 | 16,000 | Globular cluster |
| 29 | 6913 | 20 22.9 | +38 25 | Cyg | 7 | 4,000 | Open cluster |
| 30 | 7099 | 21 38.6 | −23 18 | Cap | 8 | 40,000 | Globular Cluster |
| 31 | 224 | 0 41.1 | +41 06 | And | 4 | 2,000,000 | "Great galaxy" in Andromeda |
| 32 | 221 | 0 41.1 | +40 42 | And | 9 | 2,000,000 | Elliptical companion to M31 |
| 33 | 598 | 1 32.2 | +30 30 | Tri | 7 | 3,000,000 | Spiral galaxy |
| 34 | 1039 | 2 40.1 | +42 40 | Per | 6 | 1,400 | Open cluster |
| 35 | 2168 | 6 07.0 | +24 21 | Gem | 6 | 2,800 | Open cluster |
| 36 | 1960 | 5 34.3 | +34 05 | Aur | 6 | 4,200 | Open cluster |
| 37 | 2099 | 5 50.4 | +32 33 | Aur | 6 | 4,100 | Open cluster |
| 38 | 1912 | 5 26.6 | +35 48 | Aur | 7 | 4,200 | Open cluster |
| 39 | 7092 | 21 31.1 | +48 18 | Cyg | 6 | 900 | Open cluster |
| 40 | | | | UMa | | | No cluster or nebulas? (2 stars) |

## 8. MESSIER CATALOGUE (Continued)

| M | NGC | Position (1970) R.A. | Decl. | Const. | App. Mag. | Distance (light yrs.) | Description |
|---|---|---|---|---|---|---|---|
| 41 | 2287 | 6  45.8 | −20  42 | CMa | 6 | 2,300 | Open cluster |
| 42 | 1976 | 5  33.9 | −05  24 | Ori |  | 1,500 | Great "Orion nebula" |
| 43 | 1982 | 5  34.1 | −05  18 | Ori |  | 1,500 | Associated with M42 |
| 44 | 2632 | 8  38.2 | +20  06 | Cnc | 4 | 520 | "Praesepe" (open cluster) |
| 45 |  | 3  45.7 | +24  01 | Tau |  | 400 | "Pleiades" (open cluster) |
| 46 | 2437 | 7  40.4 | −14  45 | Pup | 9 | 2,300 | Open cluster |
| 47 | 2422 | 7  35.1 | −14  26 | Pup | 5 | 1,600 | Open cluster |
| 48 | 2548 | 8  12.0 | −05  41 | Hya | 6 | 1,800 | Open cluster |
| 49 | 4472 | 12  28.3 | +08  10 | Vir | 9 | 40,000,000 | Elliptical galaxy |
| 50 | 2323 | 7  01.5 | −08  18 | Mon | 6 | 3,000 | Open cluster |
| 51 | 5194 | 13  28.6 | +47  21 | CVn | 9 | 15,000,000 | "Whirlpool galaxy" (spiral) |
| 52 | 7654 | 23  22.9 | +61  26 | Cas | 7 | 6,000 | Open cluster |
| 53 | 5024 | 13  11.5 | +18  20 | Com | 8 | 65,000 | Globular cluster |
| 54 | 6715 | 18  53.2 | −30  31 | Sgr | 8 | 55,000 | Globular cluster |
| 55 | 6809 | 19  38.1 | −31  01 | Sgr | 5 | 20,000 | Globular cluster |
| 56 | 6779 | 19  15.4 | +30  07 | Lyr | 8 | 40,000 | Globular cluster |
| 57 | 6720 | 18  52.5 | +33  00 | Lyr | 9 | 1,800 | "Ring nebula" (planetary) |
| 58 | 4579 | 12  36.2 | +11  59 | Vir | 10 | 40,000,000 | Barred spiral galaxy |
| 59 | 4621 | 12  40.5 | +11  50 | Vir | 11 | 40,000,000 | Elliptical galaxy |
| 60 | 4649 | 12  42.1 | +11  44 | Vir | 10 | 40,000,000 | Elliptical galaxy |
| 61 | 4303 | 12  20.3 | +04  39 | Vir | 10 | 40,000,000 | Spiral galaxy |
| 62 | 6266 | 16  59.3 | −30  04 | Sco | 7 | 26,000 | Globular cluster |
| 63 | 5055 | 13  14.4 | +42  11 | CVn | 10 | 16,000,000 | Spiral galaxy |
| 64 | 4826 | 12  55.2 | +21  51 | Com | 8 | 12,000,000 | Spiral galaxy |
| 65 | 3623 | 11  17.3 | +13  16 | Leo | 10 | 20,000,000 | Spiral galaxy |
| 66 | 3627 | 11  18.6 | +13  10 | Leo | 9 | 20,000,000 | Spiral galaxy |
| 67 | 2682 | 8  49.5 | +11  56 | Cnc | 7 | 27,000 | Old open cluster |
| 68 | 4590 | 12  37.8 | −26  35 | Hya | 8 | 40,000 | Globular cluster |
| 69 | 6637 | 18  29.8 | −32  23 | Sgr | 8 | 23,000 | Globular cluster |
| 70 | 6681 | 18  41.3 | −32  19 | Sgr | 9 | 65,000 | Globular cluster |
| 71 | 6838 | 19  52.4 | +18  42 | Sge | 9 | 16,000 | Globular cluster |
| 72 | 6981 | 20  51.8 | −12  41 | Aqr | 9 | 68,000 | Globular cluster |
| 73 | 6994 | 20  57.3 | −12  46 | Aqr |  |  | Open cluster |
| 74 | 628 | 1  35.1 | +15  38 | Psc | 11 | 25,000,000 | Pinwheel spiral galaxy |
| 75 | 6864 | 20  04.3 | −22  01 | Sgr | 8 | 94,000 | Globular cluster |
| 76 | 650 | 1  40.3 | +51  25 | Per | 11 | 15,000 | Planetary nebula |
| 77 | 1068 | 2  41.1 | −00  07 | Cet | 9 | 40,000,000 | Seyfert galaxy |
| 78 | 2068 | 5  45.3 | +00  02 | Ori |  | 1,600 | Small gaseous nebula |
| 79 | 1904 | 5  22.9 | −24  33 | Lep | 8 | 50,000 | Globular cluster |
| 80 | 6093 | 16  15.2 | −22  55 | Sco | 7 | 36,000 | Globular cluster |

## 8. MESSIER CATALOGUE (Continued)

| M | NGC | Position (1970) R.A. | Decl. | Const. | App. Mag. | Distance (light yrs.) | Description |
|---|---|---|---|---|---|---|---|
| 81 | 3031 | 9 53.4 | +69 12 | UMa | 8 | 7,000,000 | Spiral galaxy |
| 82 | 3034 | 9 53.6 | +69 50 | UMa | 9 | 7,000,000 | Exploding galaxy |
| 83 | 5236 | 13 35.3 | −29 43 | Hya | 9 | 15,000,000 | Spiral galaxy |
| 84 | 4374 | 12 23.6 | +13 03 | Vir | 10 | 40,000,000 | Elliptical galaxy |
| 85 | 4382 | 12 23.8 | +18 21 | Com | 10 | 40,000,000 | Elliptical galaxy |
| 86 | 4406 | 12 24.6 | +13 06 | Vir | 10 | 40,000,000 | Elliptical galaxy |
| 87 | 4486 | 12 29.2 | +12 33 | Vir | 10 | 40,000,000 | Elliptical galaxy with jet |
| 88 | 4501 | 12 30.4 | +14 35 | Com | 10 | 40,000,000 | Spiral galaxy |
| 89 | 4552 | 12 34.1 | +12 43 | Vir | 11 | 40,000,000 | Elliptical galaxy |
| 90 | 4569 | 12 35.3 | +13 19 | Vir | 11 | 40,000,000 | Spiral galaxy |
| 91 | | | | | | | (M58?) |
| 92 | 6341 | 17 16.2 | +43 11 | Her | 7 | 32,000 | Globular cluster |
| 93 | 2447 | 7 43.2 | −23 48 | Pup | 6 | 36,000 | Open cluster |
| 94 | 4736 | 12 49.6 | +41 17 | CVn | 9 | 15,000,000 | Spiral galaxy |
| 95 | 3351 | 10 42.3 | +11 52 | Leo | 10 | 25,000,000 | Barred spiral galaxy |
| 96 | 3368 | 10 45.1 | +11 59 | Leo | 10 | 25,000,000 | Spiral galaxy |
| 97 | 3587 | 11 13.1 | +55 11 | UMa | 11 | 2,600 | "Owl nebula" (planetary) |
| 98 | 4192 | 12 12.2 | +15 04 | Com | 10 | 40,000,000 | Spiral galaxy |
| 99 | 4254 | 12 17.3 | +14 35 | Com | 10 | 40,000,000 | Spiral galaxy |
| 100 | 4321 | 12 21.4 | +15 59 | Com | 10 | 40,000,000 | Spiral galaxy |
| 101 | 5457 | 14 02.1 | +54 30 | UMa | 8 | 15,000,000 | Pinwheel spiral galaxy |
| 102 | | | | | | | (M101?) |
| 103 | 581 | 1 31.2 | +60 32 | Cas | 7 | 8,000 | Open cluster |

# 9. STELLAR ATLAS

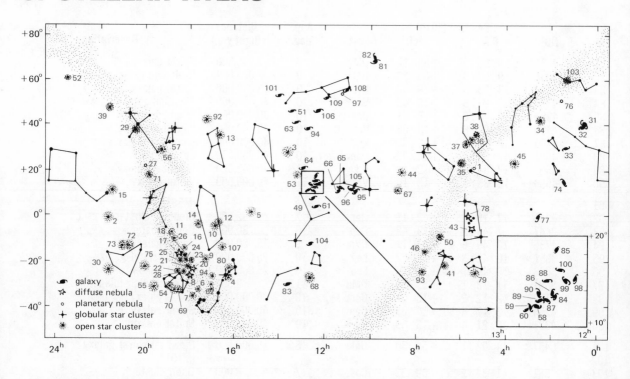

Legend:
- 🐟 galaxy
- ✩ diffuse nebula
- ○ planetary nebula
- ✛ globular star cluster
- ✺ open star cluster

444

# APPENDIX 2
# MONTHLY STAR CHARTS

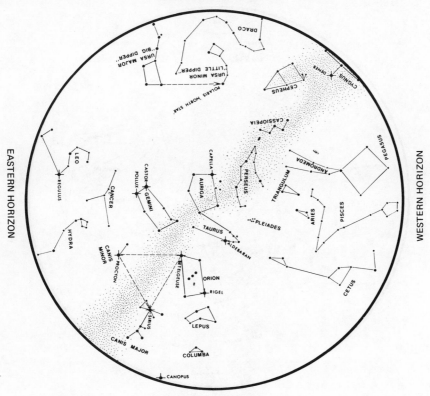

NORTHERN HORIZON

EASTERN HORIZON

WESTERN HORIZON

SOUTHERN HORIZON

## THE NIGHT SKY IN JANUARY

Latitude of chart is 34°N, but it is practical throughout the continental United States.

To use: Hold chart vertically and turn it so the direction you are facing shows at the bottom.

Chart time (Local Standard):

10 p.m. First of month
9 p.m. Middle of month
8 p.m. Last of month

Star Chart from GRIFFITH OBSERVER monthly magazine

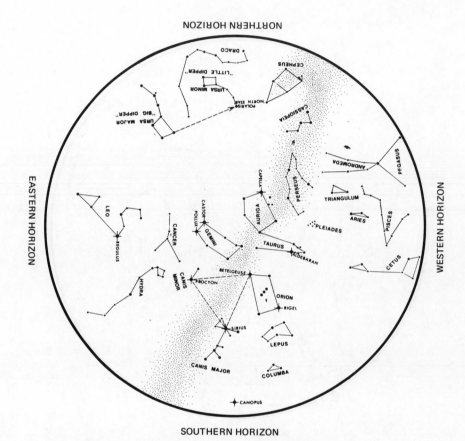

**THE NIGHT SKY IN FEBRUARY**

Latitude of chart is 34°N, but it is
practical throughout the continental
United States.

To use: Hold chart vertically and turn
it so the direction you are facing
shows at the bottom.

Chart time (Local Standard):

10 p.m. First of month
9 p.m. Middle of month
8 p.m. Last of month

Star Chart from GRIFFITH OBSERVER monthly magazine

446

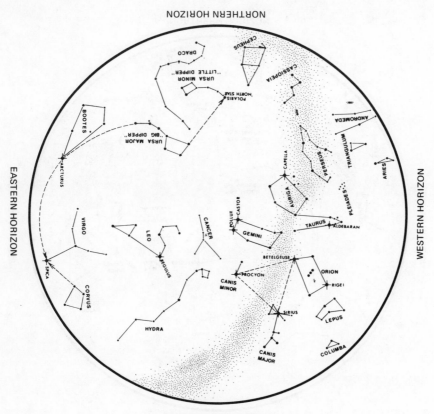

## THE NIGHT SKY IN MARCH

Latitude of chart is 34°N, but it is practical throughout the continental United States.

To use: Hold chart vertically and turn it so the direction you are facing shows at the bottom.

Chart time (Local Standard):

10 p.m. First of month
9 p.m. Middle of month
8 p.m. Last of month

Star Chart from GRIFFITH OBSERVER monthly magazine

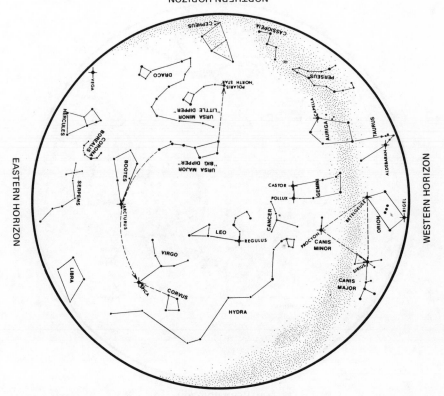

NORTHERN HORIZON

SOUTHERN HORIZON

THE NIGHT SKY IN APRIL

Latitude of chart is 34° N, but it is
practical throughout the continental
United States.

To use: Hold chart vertically and turn
it so the direction you are facing
shows at the bottom.

Chart time (Local Standard):

10 p.m. First of month
9 p.m. Middle of month
8 p.m Last of month

Star Chart from GRIFFITH OBSERVER monthly magazine

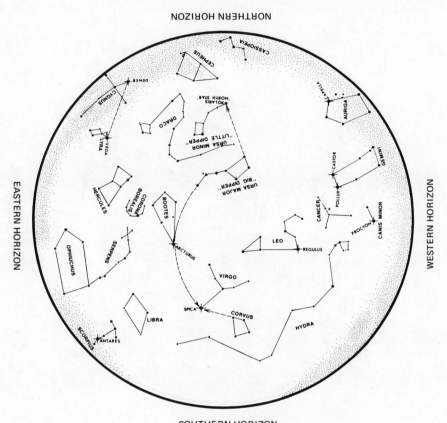

SOUTHERN HORIZON

THE NIGHT SKY IN MAY

Latitude of chart is 34°N, but it is
practical throughout the continental
United States.

To use: Hold chart vertically and turn
it so the direction you are facing
shows at the bottom.

Chart time (Local Standard):

10 p.m. First of month
9 p.m Middle of month
8 p.m. Last of month

Star Chart from GRIFFITH OBSERVER monthly magazine

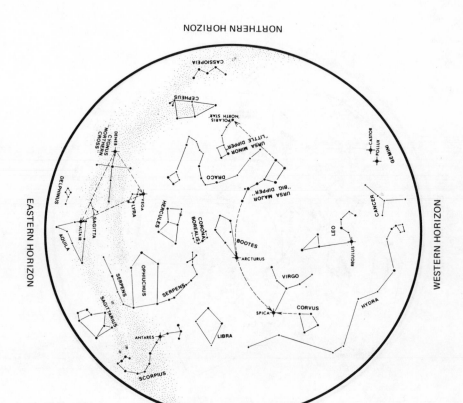

NORTHERN HORIZON

EASTERN HORIZON

WESTERN HORIZON

SOUTHERN HORIZON

## THE NIGHT SKY IN JUNE

Latitude of chart if 34°N, but it is
practical throughout the continental
United States.

To use: Hold chart vertically and turn
it so the direction you are facing
shows at the bottom.

Chart time (Local Standard):

10 p.m. First of month
9 p.m. Middle of month
8 p.m. Last of month

Star Chart from GRIFFITH OBSERVER monthly magazine

450

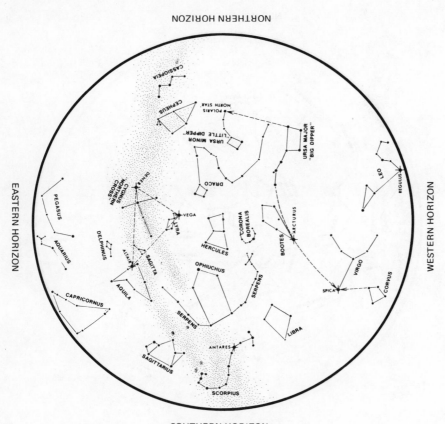

## THE NIGHT SKY IN JULY

Latitude of chart is 34° N, but it is
practical throughout the continental
United States.

To use:  Hold chart vertically and turn
it so the direction you are facing
shows at the bottom.

Chart time (Local Standard):

10 p.m. First of month
9 p.m. Middle of month
8 p.m. Last of month

Star Chart from GRIFFITH OBSERVER monthly magazine

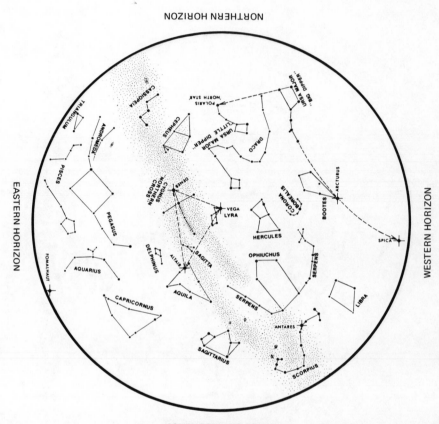

SOUTHERN HORIZON

## THE NIGHT SKY IN AUGUST

Latitude of chart if 34°N, but it is
practical throughout the continental
United States.

To use: Hold chart vertically and turn
it so the direction you are facing
shows at the bottom.

Chart time (Local Standard):

10 p.m. First of month
9 p.m. Middle of month
8 p.m. Last of month

Star Chart from GRIFFITH OBSERVER monthly magazine

NORTHERN HORIZON

EASTERN HORIZON

WESTERN HORIZON

URSA MAJOR "BIG DIPPER"

URSA MINOR "LITTLE DIPPER"

DRACO

POLARIS NORTH STAR

POLLUX

CASTOR

GEMINI

CEPHEUS

VEGA

LYRA

CYGNUS NORTHERN CROSS

CAPELLA

AURIGA

CASSIOPEIA

SAGITTA

CYGNUS

BETELGEUSE

TAURUS

PERSEUS

DELPHINUS

ALTAIR

AQUILA

ORION

ALDEBARAN

PLEIADES

TRIANGULUM

ANDROMEDA

ARIES

PEGASUS

RIGEL

PISCES

AQUARIUS

CAPRICORNUS

CETUS

FOMALHAUT

GRUS

SOUTHERN HORIZON

## THE NIGHT SKY IN SEPTEMBER

Latitude of chart is 34° N, but it is practical throughout the continental United States.

To use: Hold chart vertically and turn it so the direction you are facing shows at the bottom.

Chart time (Local Standard):

10 p.m. First of month
9 p.m. Middle of month
8 p.m. Last of month

Star Chart from GRIFFITH OBSERVER monthly magazine

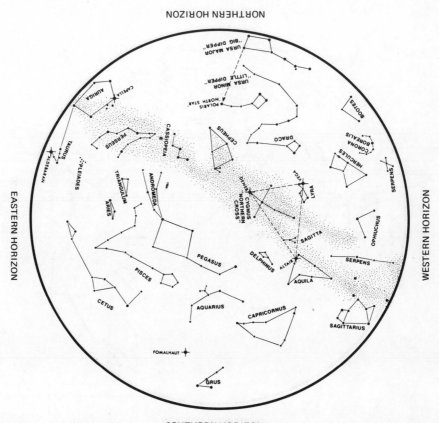

NORTHERN HORIZON

SOUTHERN HORIZON

## THE NIGHT SKY IN OCTOBER

Latitude of chart is 34°N, but it is
practical throughout the continental
United States.

To use: Hold chart vertically and turn
it so the direction you are facing
shows at the bottom.

Chart time (Local Standard):

10 p.m. First of month
9 p.m. Middle of month
8 p.m. Last of month

Star Chart from GRIFFITH OBSERVER monthly magazine

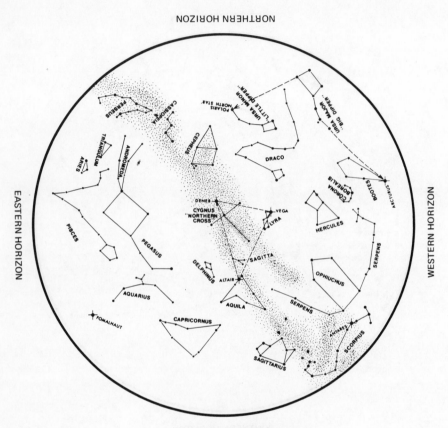

NORTHERN HORIZON

EASTERN HORIZON

WESTERN HORIZON

SOUTHERN HORIZON

## THE NIGHT SKY IN NOVEMBER

Latitude of chart is 34°N, but it is
practical throughout the continental
United States.

To use: Hold chart vertically and turn
it so the direction you are facing
shows at the bottom.

Chart time (Local Standard):

10 p.m. First of month
9 p.m. Middle of month
8 p.m. Last of month

Star Chart from GRIFFITH OBSERVER monthly magazine

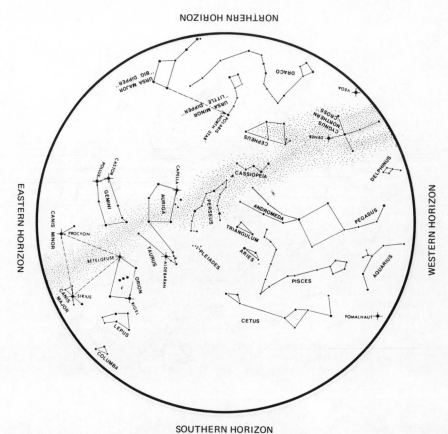

NORTHERN HORIZON

EASTERN HORIZON

WESTERN HORIZON

SOUTHERN HORIZON

## THE NIGHT SKY IN DECEMBER

Latitude of chart is 34°N, but it is
practical throughout the continental
United States.

To use: Hold chart vertically and turn
it so the direction you are facing
shows at the bottom.

Chart time (Local Standard):

10 p.m. First of month
9 p.m. Middle of month
8 p.m. Last of month

Star Chart from GRIFFITH OBSERVER monthly magazine

# INDEX

466

74 75 76   9 8 7 6 5 4 3 2 1